UFO Drawings
From the Cantonese world
廣東話世界的不明飛行物繪本

Compiled By Cheuk Fei

卓飛 編著

Since that day

When I was 13 years old, in that sultry dusk, I enjoyed the shade on the roof of the 13th floor with my father and two younger sisters. That night, let us meet this strange amber ball-shaped luminous body, which changed me.

That strange experience did not change me immediately, but instead, it planted a seed in my mind. In the stage of money, power in Hong Kong, pursuing UFOs is nonsensical. It is a lonely path that no one understands unless you witness them.

At that moment, the strange glowing sphere flying completely silent, sometimes fast and sometimes slow, obviously with autonomous power, consciously wandering around, as if looking for something. Sometimes red, and occasionally orange, permanently imprinted in mind.

At the age of 30, I was determined to leave the tiring advertising industry. The seed planted in my mind began to sprout and finally burst out in my brain. I have experienced a lot & learned a lot in 10 years; I eventually became a talk show host. And because of this capacity, witnessing stories and images from the audience become a lot more than I can imagine.

Although the place is small, our light is immense.

There is a saying, looking at the earth at night from outer space, the light emitted by the city should be charming to civilizations outside the planet. Imagine visiting another living planet, and you will also choose a place with vitality to visit. Hong Kong is a small land with beautiful night scenes that are famous in the world. "Once" was the Pearl of the East on the earth, which may explain why this small island has numerous UFO reports.

In the past few years, my work has been fortunate to be recognized in Western society. While foreign friends and researchers are interested in UFOs in China, this Cantonese-speaking international city is unexpectedly unpopular. Hong Kong has a small area on the earth. It's a small place, but its influence involves the whole world. After June 2019, Hong Kong has attracted the attention of the entire world. I believe that researchers

從那天起......

13 歲那年，那個悶熱的黃昏，和父親及兩個妹妹一起在 13 樓的天台乘涼，那個晚上，讓我們一起遇上這個奇怪橙黃色（Amber）的球型發光體，從此改變了我......

那個奇怪經歷，沒有即時改變了我，卻在我的腦中種下了一粒種子。在香港這個金錢，權力，利益的競技場上，每天的思維除了是工作外，還是工作......根本沒有任何時間去追逐一片沒有經濟效益的資訊，別人也無法理解那種完全震憾的感覺。

那一刻看著一個發光奇怪的東西，完全無聲音地飛行，時快時慢，明顯不是隨風飄浮，是有自主性的動力，有意識地徘徊，好像在尋找甚麼。那種時紅色時橙色的轉變，永久烙印在腦中。

30 歲那年，我決心離開很累人的廣告行業，靜下來時，種在腦內的這粒種子，開始萌芽，最終在腦內爆發......我要尋找精彩的人生，終於我用了 10 年經歷很多，學到很多，我當上節目主持人。亦因為這個身份，我能夠收集很多從聽眾的目擊故事，及圖像。

地方雖小，但我們光芒卻很大......

曾經有一個說法，在外太空上看夜間的地球，城市發出的光芒應該很吸引地球以外的文明。試想像一下你準備去另一個有生命的星球參觀，你也會選擇一個有生命力的地方遊覽。香港這片小小土地卻有著世界上很著名的美麗夜景，「曾經」是地球上的東方之珠，可能解釋了為何這個小島有著大量的不明飛行物報告。

過去數年，我的工作有幸在西方社會讓人認識，當外國的朋友及研究者都對中國的 UFO 感興趣的同時，這個說着廣東語的國際城市居然無人問津，香港在地球上面積小得很的彈丸之地，影響力卻涉及整個世界。2019 年 6 月以後，香港掀起了世界的注

should be deeply interested in the UFO story in this small place.

I believe that the giant Mothership that occurred in Wah Fu Estate in the 1980s was not the only sighting.

By the witness's hand, regression what was witnessed at the time, and then draw it on paper. Each picture is a seed in their mind. Every seed is full of energy, and I want them to germinate. I want these stories to spread all over the world. And to organize a book of UFOs seen by Cantonese-speaking witnesses. I want to gather the most significant number of stories and images into this book you are reading. Distribute the strange sightings of Hong Kong and Cantonese to the world.

My Talkshow "MJ13" is mainly hosted in Cantonese, and the audiences are also Cantonese speakers. Mainstream historical archaeology said that the history of Cantonese traced back to the Han Dynasty in 220 AD, and the ancient Chinese also used it. So old Chinese UFO records on this language.

I want to leave a few references in the world of UFOs with my mother language when the world's pace intends to gradually forget this ancient language. Cantonese, Traditional Chinese is a heritage of Chinese culture and cannot be forgotten.

I must leave an unforgettable portrait in the UFO in Hong Kong when the world's pace intends to forget about this place. Moreover, Hong Kong cannot be forgotten. I am proud of being a Hong Konger.

Every sighting is a record of a Cantonese-speaking person, and most sightings are in Hong Kong or shock sightings by Hong Kong people.

Cheuk Fei

目，整個世界對她無人不識，我想去相信（I want to believe），西方世界的研究者對這片小地方的不明飛行物故事應該深感興趣吧。

我想去相信，80 年代發生在華富邨的巨型母艦式（Mothership）的飛行物不是唯一的目擊……

我想去相信，利用目擊者自己的手，拿著筆回溯（regression）當時目擊的情況，再畫在紙上，每張圖都是每個目擊者腦內的種子，充滿能量，我想讓大家腦內的種子都萌芽，要讓這些故事傳到世界各地，整理一本說廣東語的目擊者所看見的不明飛行物，我要把最大量的故事、圖像集結成一份報告。一本書，把香港、廣東語的奇異目擊分享到全世界去……

我的節目（Talk show）MJ13 是以廣東語為主，聽眾亦是廣東語使用者，根據主流歷史考古學相信：廣東語的歷史可追溯至公元220年的漢代，古代的中國人也是使用廣東語，因此很多古代中國的不明飛行物記錄亦是以此語言為記錄基礎。

當世界上的步伐似乎要逐步忘卻這種古代語文，以此為母語的我卻很想在 UFO 的世界裡留下少少的參考。讓世界的人知道這裡發生過的事……廣東語、繁體中文，是中國文化的遺產，不能被遺忘。

而當世界上的步伐似乎要逐步忘卻這個獨特的香港，以此為出生地的我決意在 UFO 世界文化內留下不能忘卻的畫像。讓世界的人知道……即使這個地方以後文明不再，香港是不能被遺忘的，我以身為香港人為傲。

每一個目擊都是說廣東語的人的記錄，大部份的目擊都是在香港，或是香港人的震憾目擊。

卓飛

Mothership in Aberdeen, Hong Kong at the 1980s

Case 000

Date: Unknown (Somewhere between 1981 - 1982, Summer...)
Time: After midnight... (Somewhere between 00:00 - 02:00)
Location: Wah Fu Estate, Aberdeen, Hong Kong.
Description: The most well-known UFO case in Hong Kong is about a vast Octagon-shaped object visiting an estate in the southern part of Hong Kong Island. As per testimony, multiple witnesses saw the giant object, so it was a mass sighting case.

Kwan (nickname) was a teenager when this incident occurred. He awoke by some weird deep, humming voice that shook the whole building around midnight. He realized something unusual on the street when he walked past the window.

"The street is darker than usual!"

"The street is darker than usual!" He thinks. And then he realized that people who live in the opposite building are also awake, and quite a lot of people are learning and looking out the window. As most people should be in bed at night, a pretty lighted-up building is quite usual. Then he realized the people were staring at the roof of the building where they were living. And people are looking at the top, staring and discussing. "Somewhere weird happening," he thinks.

He woke his younger brother up and went out to the public area of the elevator to have a better view of their building's roof. They saw a vast black octagon or hexagon-shaped object with a very shaped edge hovering on the top of the building where they were living. The street is darker than usual because the soccer field-sized object blocks the moonlight and starlight.

People around that area are curious; some are yelling, some are panicking,
"What's going on?"
"Guanyin* is coming!"
"What the hell is it?"
"The War is coming!!"

After some short moment, the massive black object created some panicking when it shook one last humming blast; Kwan said he remembered seeing a very bright luminous blue glow shined at the bottom side of the craft. It slowly moved away from the building and showed its blue light. Floated to the open sea area and vanished into the night sky. The whole sighting from Kwan is about 15-20 minutes.

"Everyone talks about it the next morning!" Kwan said. And there were some university students recorded to the people outside the elevator and bus terminal within the estate. The topic was discussed for weeks in the restaurant and bus stations.

It has no official record, and UFO information was not popular in those days. However, many neighbors saw the UFO in Kwan's testimony; only a few were willing to admit the sighting.

Second case?

Anita was a teenager when she saw an enormous octagon shape object hovering on top of a building in the same estate named Wah Fu Estate*. But the time was different from Kwan's. Instead of midnight, it happened at around 17:00-18:00. Her balcony was facing the bus terminal, and it was Off work hours, and the bus terminal was busy.

Anita's mother told her to close the cover lid because of the strong wind. When she was about to push the plastic lid covering the balcony's big window, she saw a massive object hovering in the sky. It was shining and beautiful with many colorful lights at the bottom—its description quite similar to the one that Kwan observed.

Interestingly, no one on the street acknowledges the mystery object because everyone is rushing to return home for dinner after a hard day of work. It flew and disappeared after a short moment.

I believe there were at least two visits to the southern side of the island by this giant octagon mothership. And a lot smaller UAP in that particular area because of the open space of the ocean. People intend not to look up at the sky in the crowded city, which explains why Anita is the only witness in the second case described above. The UFO probably caused the strong wind mentioned in the story.

*Guanyin - Goddess of Mercy in Buddhism & Eastern religions.
*Wah Fu Estate - Public housing estate located in Hong Kong's Southern District, built in 1967.

華富村 UFO 事件

相信如果你是香港人，或是筆者的聽眾，應該對「80年代華富村集體目擊」不會陌生，這個案亦是我決心出這本書的背後原因之一，當我遊走西方國家時，發現這個對於香港人有着重大影響的個案，居然無人知道，思前想後多年發現這完全是語言問題。

因此我在這書用頗多的內容介紹給外國的讀者，亦因此，我把這個案列入第 000 號個案。

大約 1980 年代初期，年份大約 1981-82（後來目擊證人估計）凌晨時份，當年十多歲的阿君被屋內傳出的低頻震動吵醒，正準備行去洗手間，往窗外看時，發現所看的景像有點不能解釋的異常。

「街道比平常時更暗！」不知道何解街道看起來好像很暗的樣子，即使街燈是亮著的，整個位置很深色，平常即使是夜深，街仍然很光亮。

還有一點不尋常是，對面大廈的住戶很多都亮著燈，而且很多住客也靠著窗看著阿君住的華泰樓。十多歲的阿君感覺這個現像有點奇怪，心想一定有些不尋常的事發生了，他再靜心觀察下，發現對面大廈的住客都是看著自己所住的大廈的天台，而且還聽到這些住客在大叫，有人喊道：對面搞甚麼？有人喊出：打到來啦！！！甚至又些更大叫：觀音顯靈呀！！

阿君即時叫醒仍熟睡的弟弟，他還嘗試叫醒母親，可惜不果。兩兄弟馬上跑到走廊，並向他們居住那層數的電梯位置跑去，因為那個位置有一個大空間可以看到天台。當他們到達後往天台一看，他們看見一大片黑色的物件在天台向外伸展出去，而該物件很大，完全把夜空的星光或者雲層的反光遮擋著，形成一個巨型的黑影，這就解釋為何街上這麼暗。

接著不久，這件巨大的遮蔽物慢慢開始移動，並且發出冷藍色的光，現像一片震憾，有人大叫，而這隻物件慢慢飄離華泰樓的上空，這時，阿君及弟弟才發現這是一隻六角型或是八角型的巨大飛行物，它慢慢向著海的方向飛走，整個目擊歷時大約15-20分鐘。

根據阿君所述，第二天早上途人皆議論紛紛，有些大學學生嘗試在電睇大堂和住客做訪問記錄，由於當年香港人對 UFO 這個現像頗為陌生，因此亦沒有正式記錄，但傳媒多年來找到數名目擊者願意承認及提供資料，阿君只是其中一位。

華富村事件其實還有另一個個案，Anita是一名學生，大約亦是80年代她在家亦見過，不過時間卻跟阿君的目擊時間不同，當時為黃昏「放工時間」，Anita 的家向著華富村內的巴士總站，當時突然刮起大風，Anita 被叮囑把擋風板關上時，看見天上有一隻巨型的物件停著，其底部發出彩色美麗的光，奇怪的是當年沒有其他人注意到奇怪飛行物的存在，好像每個人也低著頭趕回家一樣。不久後巨型彩光飛行物亦飛走了。

華富村其後亦多次有人目擊大大小小的不明飛行物。有人說是因為該區為飛行航道，有人說對開海底可能是「飛碟基地」，我覺得最主要原因是該區比較罕有地「開揚」，看見天空的位置比較多，因此人們望向天的機會也多。大都市的人都是不望天，現代大都市人更只會看著手機吧。

My Own Sighting

Like the other days, my father routinely brought me and
two of my younger sisters went up to the rooftop of the
building we lived. It was summer, and not everybody could
afford an air-conditioner back in those days. We love the
chill air at night when the sun is down, and the moon is
up. While my father was reading his newspaper, my sisters
played with their little dolls made of paper. I saw an amber
light shining in the sky at the opposite hillside. The hill was
called "Mount Parker." It was around 200 feet away from
our rooftop. And the weird light hovering 50 feet up of
the tree lines. It was amber in color but growing red light.
It starts making circular flying patterns, sometimes faster,
sometimes slowing down, and even stopping for a few
seconds. I have a feeling that it is searching for something.
Suddenly, it flys toward us, which made me scared. I
grabbed my father's arm and wanted to leave, and he
told me not to be afraid and to stay and watch for more.
Surprisingly. the object stopped closing to us and started
going. I feel like it can sense my fear and doesn't want
to scare me more. It flies away and disappears behind the
mountain.
When it got close, I could see something moving inside
the yellow sphere. The best way to describe it is when you
observe a stirring coffee with cream in it, the cream will
spiral, but the spiraling has consciousness.

　　13歲那年，那天的晚上，父親帶著我和兩個妹
妹，像往時一樣走上我們居住的大廈頂樓，我們差不

多每天也會這樣做。由於年少時家裡沒有冷氣 （ 當年並不是每家也能買得起冷氣機，所以夏天最好就是跑到天台乘涼。 ）

　　如往常一樣，晚飯後大約 7:30 左右，我們在大廈的頂樓天台，父親在閱讀報紙，妹妹們在玩她們的小玩具，而我則無聊四處踢。

　　突然間，我們看見對面的「柏架山」上有一件奇怪的物件停在天上，跟我們的距離大約一百至二百呎，距離樹頂五十多呎高的位置，停了一粒橙色，但發出紅色光的物體。它以圓環形的路線在慢慢移動，時快時慢，不停在轉圈，好像是在找尋甚麼似的。

　　突然看件它有點向我們的方向飛來，我感到害怕，拉著父親的手想走，父親卻說：見到這樣奇怪的東西當然要看清楚吧！！突然該飛行物好像知道我很害怕，即時沒有再飛向我們，慢慢飛到山後，並且消失於視線範圍。整個過程完全沒有發出聲音。

　　當它飛得近時，我看見這個黃色的物件，好像一個玻璃球，中間勉強看到有一些陰影在郁動，好像你在攪拌一杯咖啡，裡面有呈螺旋狀的奶油，而那個螺旋的轉動是有意識的，不是物理性的。

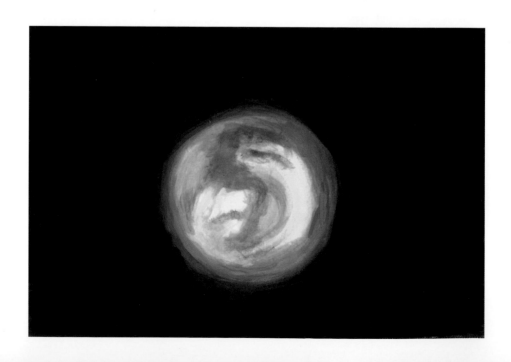

Fastwalker

UFO 記錄 1R

← 含景在左邊天空

● ←奇異的「影」

Case 001

Date: December 2008
Time: After 18:00
Location: Discovery Bay, Hong Kong.
Description: A star next to Venus suddenly dropped down to the horizon and disappeared.

已忘記了確實日期，只記得是「哈哈笑」天象後數天，翻查記錄應該是 2008 年 12 月。
當時，在愉景灣工作的我仍希望看到笑臉，所以連續幾晚 6 時下班後，依然會抬頭仰望黑
漆的夜空。可是當晚笑臉不再了，但金星很明亮；在欣賞之時發現右邊天空較遠處有一顆
與金星相同亮度與大小的星。雖然不熟悉星體名字，但這亮度的星是少見的。正想多看幾
眼，那顆星突然迅速向下移動，可以說是墜落......是一種不可能發生的速度，直至在落在
遠處水平線消失了。我馬上向著那方向跑至山坡邊，但沒法再看到這奇異的「星」。

Case 002

Date: July 2018
Time: After sunset
Location: Sedona, Arizona, USA.
Description: a star-like object passed by the sky, withness point the object by a laser pointer, object flashed twice before disappeared.

2018 年，七月一個晚上在美國亞歷桑拿州，塞多納一山頭上夜觀天空，這是我真正直接感受到會同天空上的不明飛行物體有接觸的感覺。Miss Melinda Leslie 會教我們如何分別在天空上的不明飛行物體光點、飛機和衛星信號燈。當確定光點後，就要重複做以下程序，用手上的激光電筒直接照射光點開關三次後，大叫 Power Up Please！奇妙地，光點會發出比平常光幾倍的閃光回應一至兩次後，就飛走消失於天空下。在畫上記錄一光點由發現至消失，他最少閃光 17 次地回應我們，真是難以置信，但是我不能不相信！！

周先生 2020

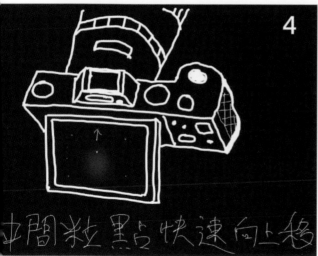

中間粒黑點快速向上移

Case 003

Date: 23 July 2021
Time: Around 20:00
Location: Sunny Bay, Hong Kong.
Description: A star-like object moving faster than the plane makes a sudden turn on first appearing, fast-moving strict line on second appearing.

7 月 23 號夜晚大約 8 點左右我同朋友喺欣澳站對出石灘向屯門方向影 f3 neowise 慧星。朋友用相機放大搵慧星時發現有一粒光點移動比飛機快。出現兩次，第一次出現睇到粒光點高速轉彎但係嚟唔切影低。第二次有用手機影低。

Lights

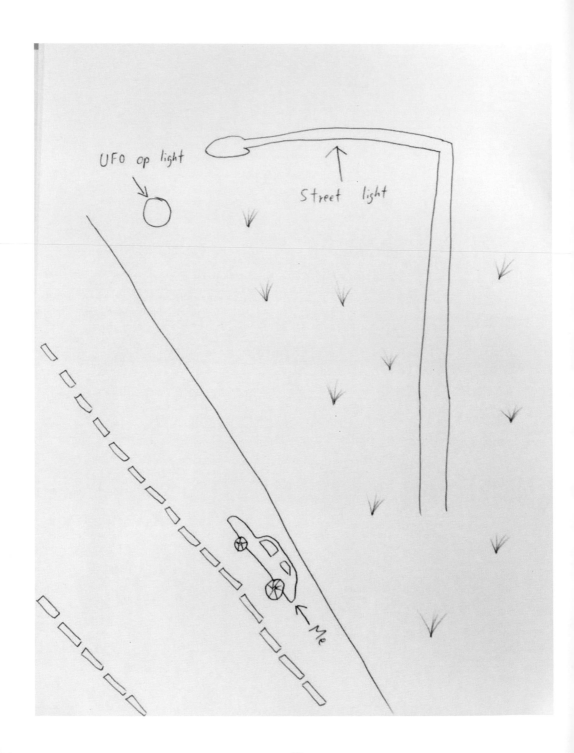

UFO op light

Street light

←Me

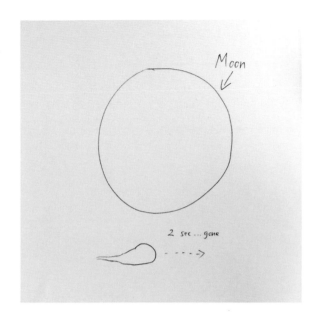

Case 004

Date: Unknown
Time: Unknown
Location: Unknown
Description: A teardrop-shaped object moved very fast under the moon and disappeared in two seconds.

日期：不明
時間：不明
地點：不明
一個淚珠狀的物體在月光下移動得非常快，兩秒鐘後就消失了。

Case 005

Date: July / August 2018
Time: 16:00-17:00
Location: Kwun Tong Highway, Kowloon, Hong Kong.
Description: A very bright metal object stays high in the sky for three minutes.

事件喺 2018 年唔記得七月定八月，下午 4、5 點左右，本人搭小巴由九龍灣去秀茂坪，經觀塘繞道，本人坐在小巴嘅右邊窗口位，望向北角方向，就發現有粒好光好光嘅一粒光點（如圖左面大廈頂），本人一直望住，九龍灣望到觀塘直至望唔到為止，大約三分鐘左右，嗰粒光點都冇郁過。
由於當日係好好天萬里無雲，我睇到嗰粒光點好似金屬定鏡咁樣反光。好光好光。
大約係咁上下😄

Case 006

Date: Unknown
Time: After midnight
Location: Phuket, Thailand.
Description: A object with bright yellow light passed the forest down below the hotel.

有一年在泰國布吉酒店遇疑似不明物體個案，我在凌晨酒店望落樓下樹林有一個發光體打橫行，如果是火球咁樹林都著晒火，但至今仍然不肯定。😄

Case 007

Date: Summer 1996
Time: Unknown
Location: Tsuen Wan (Cheung Shan Estate), Hong Kong.
Description: It was about to rain, bright light in cloud. Then oval-shaped object appeared beside the cloud. Withness watched the event that lasted seven to ten minutes.

時間：1996／夏天

荃灣象山邨秀山樓面向馬路山上，在陰天準備下雨的天氣，在烏雲雷中見到有光，光係白色穿過雲，但當時以為係準備下雨的閃電。當時正在露台望向外，光和烏雲發現有疑似一團光（2個位置）分別像在雲中穿插，它們是 ⌒ ←❳50 ←像這線形的移動，因此當時肯定不是飛機，在它們移動雲旁邊我見到它原來長相像上圖，當時觀察了 7-10 分鐘。

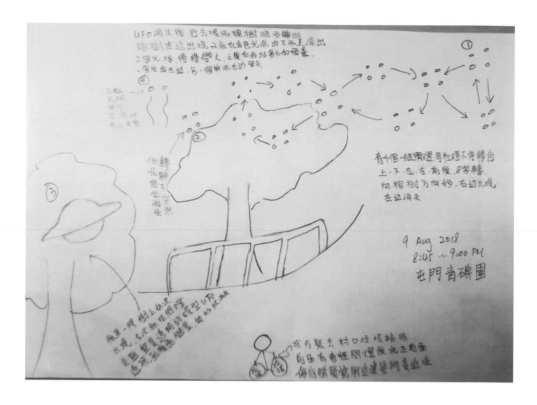

Case 008

Date: 9 August 2018
Time: 20:45 - 21:00
Location: Tuen Mun (Tsing Chuen Wai), Hong Kong.
Description: A formation by four flashing lights flying in all directions and making an infinity shape turn. Disappeared on one tree and reappear on another tree.
The object seems translucent. Two golden or white bright lights appeared again on the left side of the tree. They turned big slowly and flew in the direction of the witness, one passed by on the left side of the witness, another one passed by on the right side.

9 Aug 2018
8:45 - 9:00pm
屯門青磚圍

（圖下）我行緊去村口垃圾站時，前面有奇怪閃燈喺我正前面，停低睇發覺附近建築物變近咗

有 4 個一組閃燈，每粒燈不停轉色，上、下、左、右、前後，8 字轉，向榕樹方向移，右邊出現，左邊消失。
轉轉下突然喺呢個位消失
喺另一棵樹上再次出現，今次無咗閃燈。見到整隻透明的碟型 UFO 透視到後面樹葉，幾秒就消失
UFO 消失後行去呢兩棵樹睇多陣時，榕樹左邊出現 2 點白金色光點由下而上漂出，2 個光球慢慢變大，之後向我站著方向飛嚟，一個喺我左邊，另一個喺我右邊飛走

Case 009

Date: 8 March 2019
Time: Unknown
Location: Tuen Mun (Tsing Chuen Wai), Hong Kong.
Description: A red light object with flashing golden/white light. Passed by a light pole on the highway, make 90 degrees turn up to the sky, station in the sky and keep flashing, then disappeared when it lower to ground. Again it reappeared in the sky and lowered to the ground twice. It reappears again on top of houses.

8 March 2019
屯門藍地青磚圍

1.紅色光點下方不停閃白金橙色燈光，在西部公路燈柱上面快速飛向山邊，突然 90 度升上
  天空，停下不斷閃燈，降到山腰位消失之後，再在天空中相同位置出現再下降，重覆 2 次
2.紅色光點在山腰消失後，再見到喺村屋頂上面出現飛過。

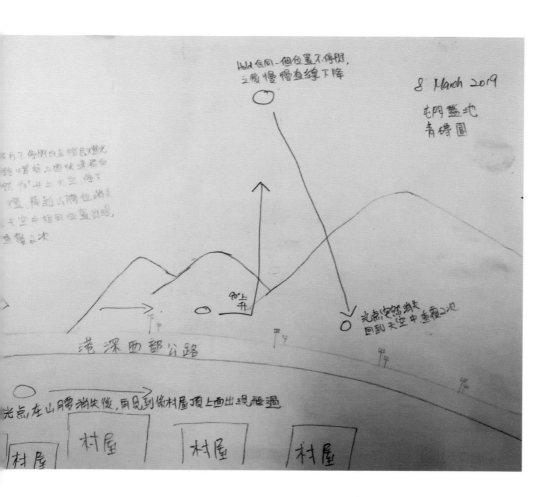

Hold 在同一個位置不停留,
之後慢慢直線下降

8 March 2019

屯門藍地
青磚圍

為了停留白色橙色閃光
於此置起上面快速罩白
光 升上天空停下
閃, 降到山腳位消失
天空中相同位置出現
重複 2次

90°上升

光點突然打大
回到天空中重複 2次

港深西部公路

光點在山腰消失後,再見到從村屋頂上面出現飛過。

村屋   村屋   村屋   村屋

33

Case 010

Date: Summer 1985
Time: 19:00 - 19:30
Location: Sau Mau Ping, Hong Kong.
Description: A giant red sphere object fly by the sky. Three witnesses mention fire-like objects surrounding the sphere and disappearing behind buildings.

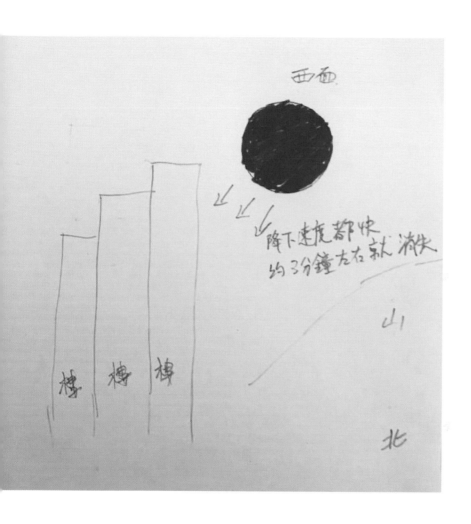

西面

降下速度都快
約3分鐘左右就消失

山

樓 樓 樓

北

36 年前某一晚,時間約 1900-1930 左右,夏天,當時天色已黑。地點秀茂坪 32 座 15 樓一單位內,在窗向外望。當時見到一大圓球,感覺好似一大火球,見到表面有像火嘅流動,依照箭咀由上空慢慢降下直到嗰 3 座樓遮住之後消失。初初我仲以為係月亮,諗下冇理由會咁,同場有 3 人睇到。

Case 011

Date: 26 September 2015
Time: 20:00 - 21:00
Location: Sha Tin, Hong Kong.
Description: A star-like object suddenly grows and slowly dims down the light until it disappears. The witness claimed to have an interactive feeling with the object.

2015 年 9 月 26 日 夜 20:00-21:00 某一晚（星期六）
當晚在沙田某住宅天台燒烤，突然有感應，望往天空，有一點特別光，感覺上好似怪怪地。一路望住佢，光度突然加強，感覺上好似開咗引擎，由光到暗跟住消失，俾我感覺由近到遠然後消失，個個望到目定口呆。

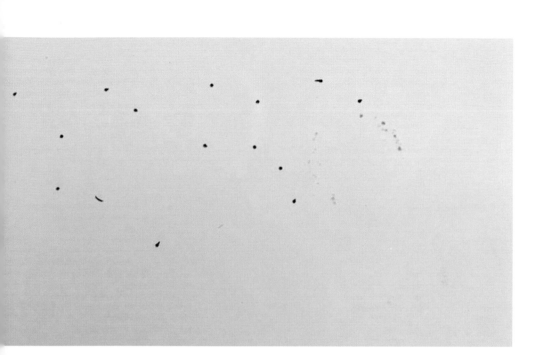

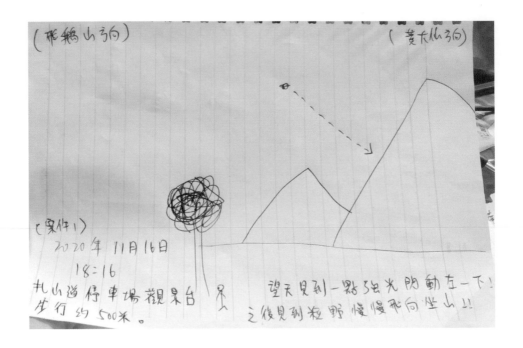

Case 012

Date: 16 November 2020
Time: 18:16
Location: Jat's Incline, Clear Water Bay, Hong Kong.
Description: An object flashed, then slowly flew in the direction of the mountain.

2020 年 11 月 16 日 18:16
札山道停車場觀景台，步行約 500 米。
望天見到一點強光閃動咗一下！之後見到粒嘢慢慢飛向座山！！

Case 013

Date: 2016
Time: Nighttime
Location: Unknown
Description: A formation of six light objects flew in the sky.

你好，我係 mj13 嘅聽眾，喺四年前嘅晚上，我上床睡覺合埋眼時想起羅茲威爾單案，之後 feel 到有一下強光，之後再望出窗，見到六顆兩排光點喺天空上飛，之後越飛越遠，消失咗。

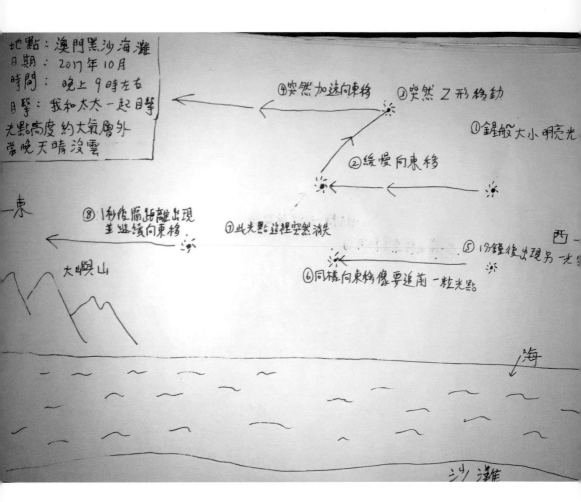

Case 014

Date: October 2017
Time: Around 21:00
Location: Praia de Hac S6, Macau.
Description: It was a clear night, and the witness saw the object with his wife.
A bright star-like object that sized like Venus moved in the east direction. Make a sudden zip-zak turn and accelerate to the east. Another object appears a minute later and chases the first object, and the second object suddenly disappears and reappears a second after.

地點：澳門黑沙海灘

日期：2017 年 10 月

時間：晚上 9 時左右

目擊：我和太太一起目擊，光點高度約大氣層外。當晚天晴無雲。

1. 金星般大小明亮光點

2. 緩慢向東移

3. 突然 Z 形移動

4. 突然加速向東移

5. 1 分鐘後出現另一光點

6. 同樣向東移像要追前一粒光點

7. 此光點這裡突然消失

8. 1 秒後隔距離出現並繼續向東移

Case 015

Date: 31 December 2019
Time: Unknown
Location: Tsz Shan Monastery (Tai Po), Hong Kong.
Description: Two formation lights flashing, suddenly flew away with high speed.

2019/12/31

慈山寺元旦叩鐘法會

在繞寺時，見天空上 2 粒星並列（覺得奇怪），右邊的一閃一閃比左邊的光亮，所以吸引
目光注意。正想跟家姐講時，突然很高速地向上飛走。

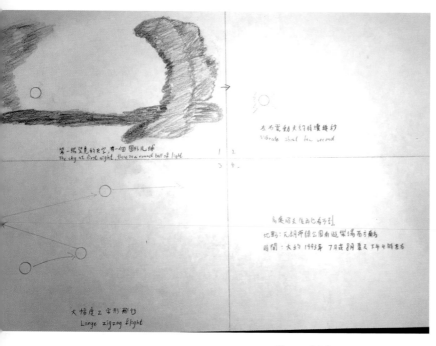

Case 016

Date: July / August 1993
Time: Around 16:00
Location: Yuen Long, Hong Kong.
Description: A bright sphere in the sky vibrated for about a few seconds, followed by a sizeable zig-zag flying, then flew away and disappeared at high speed.

地點：元朗市鎮公園南遊樂場面向南方
時間：大約 1993 年，7 月或 8 月夏天下午 4 時左右

1. 第一眼望見的天空，有一個圓形光球
   The sky at first sight, there is a round ball of light.

2. 左右震動大約持續幾秒
   Vibrate about few second

3. 大幅度 Z 字形飛行
   Large zigzag flight

4. 高速飛走後再也看不到

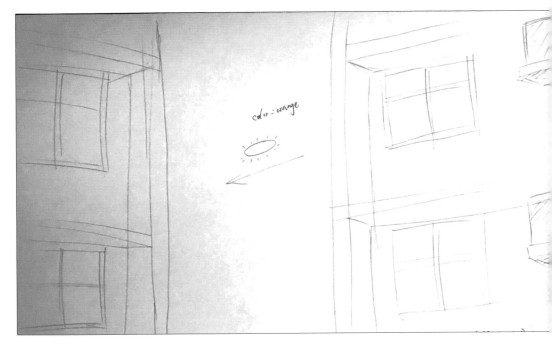

Case 017

Date: 2013
Time: 01:00
Location: Quarry Bay (Yik Cheong Building), Hong Kong.
Description: An oval-shaped object with orange light slides between buildings. Withness ensured it was neither a drone nor an airplane.

鰂魚涌益昌大廈高層 A 頭單位，望向康惠花園，兩座中間滑過橢圓形橙色發光不明飛行物體，2013 年左右，當晚大概零晨 1 點左右。肯定不是航拍機，也不是飛機。

阿力

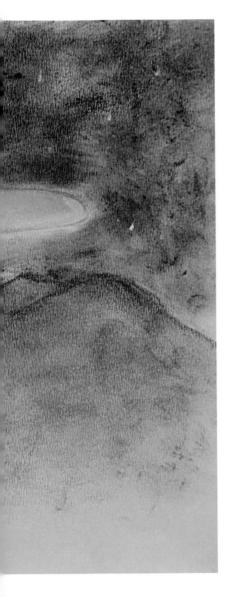

Case 018

Date: Around 1973 - 1975, summer
Time: Nighttime, unknown
Location: Tuen Mun (Nai Wai), Hong Kong.
Description: A group of primary students witnessed a huge oval-shaped object hovering in the sky. Object bright enough to light up the ground from a distance. Then a group of martial art masters saw the object while practicing Kung Fu nearby. All mentioned the object was massive. And they use the term "Mothership" to describe the object.

地點：屯門泥圍
年份：1973-75（由於相隔太耐記唔到準確日期），夏天，夜晚天空有星星，有月亮好似係滿月（方向喺另一邊）
UFO 描述：巨大鵝蛋形，整體發光，無聲

卓飛我之前都有 msg 過你我老公呢件事。
我老公當時係一個小學生，當晚一大班細路係圍村空地跟師傅學功夫。突然空地被照光佢望向圓頭山上空見到一個巨大鵝蛋形全個發光 UFO，UFO 係無聲嘅。圓頭山與空地有一段距離，但 UFO 光線足以照亮整個空地，當年圍村比較少街燈所以突然有咁光嘅嘢自然吸引到佢地成班細路注意，不過由於當年 UFO 資訊比較少而且圍村思想守舊，個個淨係識得望住個舊嘢絕對無思考過係啲咩，而且師傅要佢地繼續練拳所以就無再理會，過咗無耐就消失咗，但點消失佢就唔知。
根據圓頭山同 UFO 比例佢覺得呢個 UFO size 係 mothership。

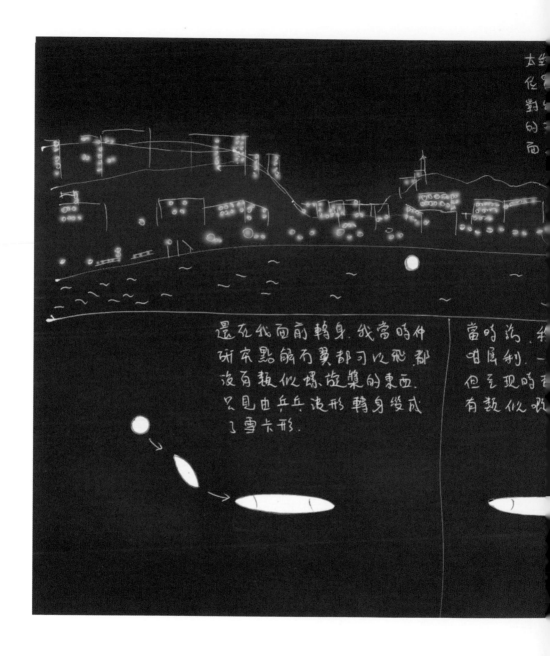

還在我面前轉身. 我當時仲
研究點解不變都咁似飛碟
沒有類似螺旋槳的東西.
只見由乒乓波形轉身後成
了雪卡形.

當時為 ...
咁屋利. 一 ...
但到現時 ...
有類似 ...

Case 019

Date: Around November 2000
Time: Around 19:00
Location: Ma On Shan, Hong Kong.
Description: A ball-shaped object flying above the sea surface, then shapeshifted to a cigar shape. Then fly away.

（上）大約在 2017 年十一月左右，時間大約晚上七點，位置向馬鞍山海濱長廊，望向科學園位置對出海面，發現在一個大約有乒乓波大小的球體，就在我直望嘅高度在海面上飛，越飛越近。

（左下）還在我面前轉身，我當時仲研究點解冇翼都可以飛，都沒有類似螺旋槳的東西，只見由乒乓波形轉身變成了雪卡形。

（右下）當時諗，而家啲科技咁犀利，一條長條形都可以飛！但至現時都沒有見過市面上有類似嘅飛行產品。

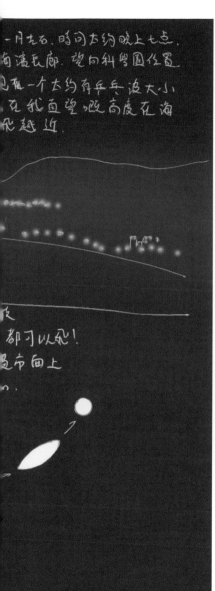

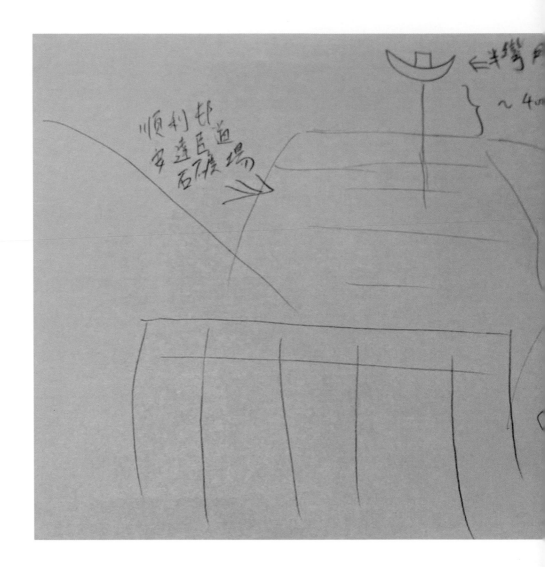

Case 020

Date: Unknown
Time: Unknown
Location: Anderson Road Quarry (Kwun Tong), Hong Kong.
Description: A crescent moon-shaped object flew in Jordan's direction.

（左）順利邨安達臣道石礦場
（中）半灣月型 UFO（發光）
（右）佐敦谷水壩

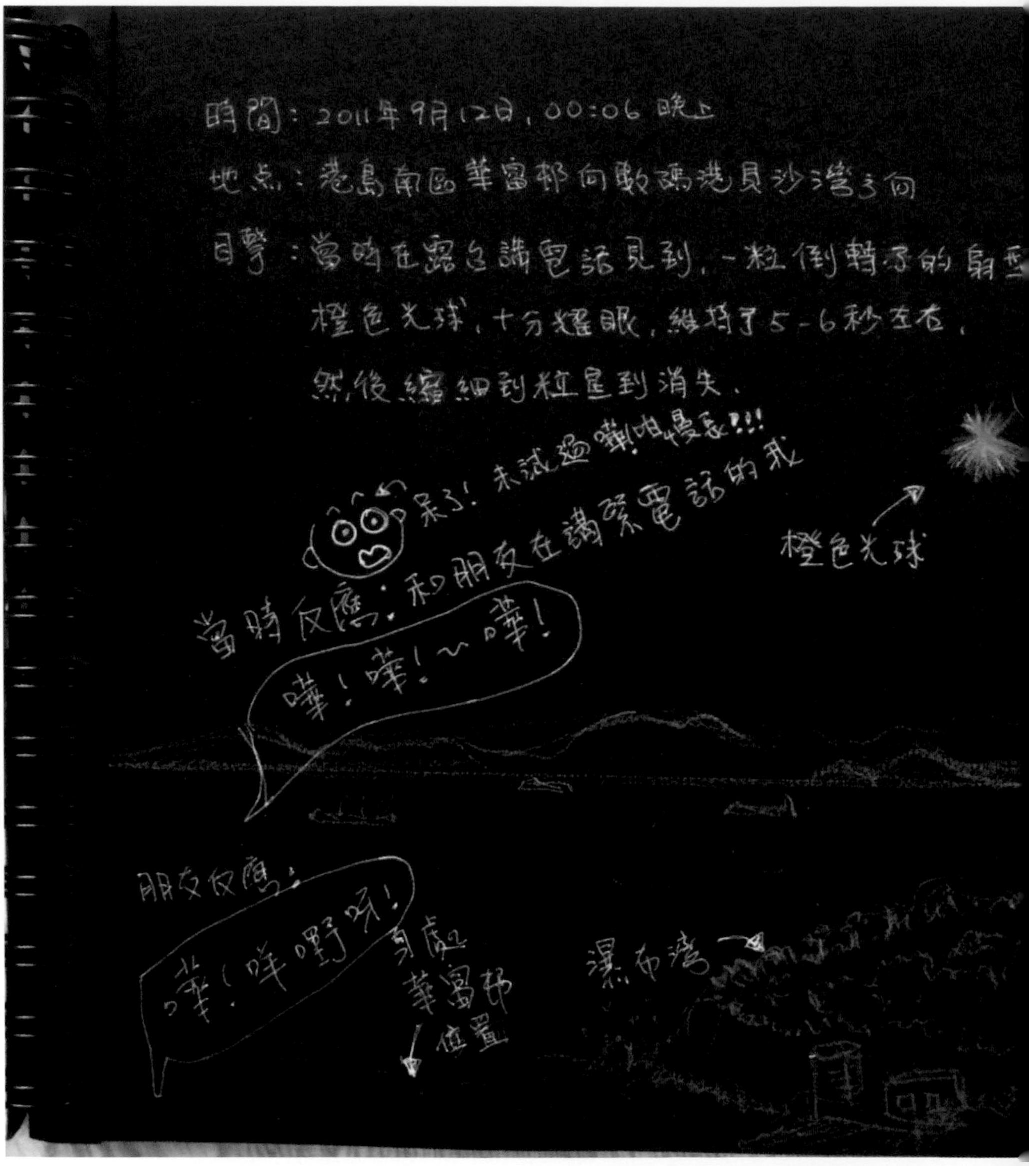

時間：2011年9月12日，00:06 晚上
地点：港島南區華富邨向數碼港員沙灣方向
目擊：當時在露台講電話見到，一粒倒轉字母J型
橙色光球，十分耀眼，維持了5-6秒左右，
然後縮細到粒塵到消失。

呆了！未試過嘞咁慢氣?!!!

橙色光球

當時反應：和朋友在講緊電話的我
嘩！嘩！～嘩！

朋友反應：
嘩！咩野呀！

身處
華富邨
位置

瀑布灣

Case 021

Date: 12 September 2011
Time: 00:06
Location: Aberdeen (Wa Fu Estate), Hong Kong.
Description: An bright object hovering in the sky was described as very bright orange light and up-side-down fan-shaped. Then the object shrunk into a star in five to six minutes and then disappeared in the dark.

時間：2011 年 9 月 12 日，00:06 晚上
地點：港島南區華富邨向數碼港貝沙灣方向
目擊：當時在露台講電話見到，一粒倒轉了的扇
　　　型，橙色光球，十分耀眼，維持了 5-6 秒
　　　左右，然後縮細到粒星消失。
當時反應：和朋友在講電話的我呆了！未試過
　　　　　嘩！咁漫長！！！

53

Case 022

Date: Second half of 2019
Time: 19:00 - 20:00
Location: Kowloon Bay, Hong Kong.
Description: It was a cloudy night, with a breeze wind. A moon-like sphere appears and then disappears without a trace. Witness confirmed it was not the moon because of the size. Then witness checked on the calendar and ensure it was not the moon because it should be the crescent phase of the moon that night.

年，月，日：2019 年下旬 (正確日子唔記得，sorry)
時間：晚上 7 點至 8 點左右
地點：九龍灣 (分隔麗晶花園 + 啓業村之間的馬路上空)
天氣情況：陰天，有雲，但雲唔算厚，而且有風。所以雲行得快 (快嚟快走)
P.S. 之後有 check 過月曆，當日係「新月」期，「月光」形狀應為「娥眉月」。

當時剛剛落巴士，準備行返屋企。因為有望天嘅習慣，所以落車後好多時都會邊行邊望天。當晚收工坐車回家，落車如常望天，在馬路邊到轉燈時發現天空有個淡黃色，球體，起動誤以為「月亮」，但因為「光球」同平時「月亮」大細及比例有出入，所以否定咗，又懷疑過係咪「飛行服務隊」直昇機，但當時無聽見直昇機聲，所以又否定。
當我過馬路到中間「安全島」位置，剛好有雲飄到擋住，到我差不多行到過完馬路時，雲層散開，同時「光球」已經完全消失。我自己覺得好唔對路，就周圍再望下天空同其他路人，想望下有無眼花，當然最後咩都無發現，連「光球」都消失咗。

*「發光球體」比例：(由地面望上天) 約 1 個 28cm 鑊大細

知知港，
森美兰州，
马来西亚

像肢林

Case 023

Date: Between 1997 - 1998
Time: 03:00 - 04:00
Location: Titi, Negeri Sembilan, Malaysia.
Description: One devil tree worker witnesses a sphere with amber light hovering in the forest, then flies in her direction. The scared woman screamed for help when her husband worked on another side of the mountain, and she claimed the amber sphere flying back and forth to herself and scared her to death. Because of too frightened, she doesn't mention the duration of the event. But she also said that people from another village met the same sphere some days ago.

這件事是發生在我舅母身上，在 1997 至 98 年間（具體記得大概這個年份），大約凌晨的 3 點多至 4 點左右的時間，鄉下的割膠工人都在這段時間出門往膠林裡工作了。那天，舅母和舅父一如既往的結伴去膠林工作。凌晨的膠林，伸手不見五指，虫鳴鳥叫，行走只靠著帶在額頭前的照燈來辨識方向。分配好工作後，舅母和舅父就各自走往負責的膠林段去開工，舅父就往高段的膠林坡上走，舅母就往低下地段走去。在舅母開始割了幾棵膠樹後沒多久，突然她看見離她不遠處有一顆發著黃燈的球體向著她的方向懸浮著飄來。因為黑暗的膠林，這被舅母認為鬼火的光球的光亮更加明亮和顯眼。據舅母說，這光亮不怎麼刺眼。當這個球體來到舅母面前（根據舅母敍述，距離少於 30-50cm）時，舅母嚇得跌坐在泥地上，連滾帶爬的想要拔腿逃離這個東西，但是被嚇著的身體根本使不到力氣，只能在地上無力的喊著舅父的名字（舅父在另一個山頭）。在這時候，這顆球還繞著舅母身邊時近時遠的靠近她。也不知道過了多久，這顆懸浮的球體就飛走了（因為被嚇著，沒能具體說出球體逗留時間）。根據舅母的敍述，附近另一條村的村民也在我舅母遇見前一些天遇到過這個球體。

* 這些遇見球體的人都認為那是鬼火......

Case 024

Date: Autumn 2016
Time: After midnight
Location: Milton Keynes, UK.
Description: An silent object hovering in the sky, then two helicopters arrived and pointed their spot-light at it... (Never mention how it disappeared or what happened next.)

時間：大約 2016 年秋季深夜
地點：英國 Milton Keynes（米頓堅斯市）
喺一個漆黑嘅夜晚，見到有個靜止的光體，周圍隱約聽見直升機聲，有兩三架直升機用光
照住中間嗰個光體......

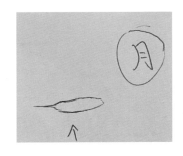

Case 025

Date: Around 2017

Time: 00:00 - 01:00

Location: Tuen Mun (Tin King Estate), Hong Kong.

Description: A teenager saw a translucent circle fly by the moon. When they overlap, the moon behind them can be seen. Then the strange circle vanished in the dark sky. A "Cartoon styled laser blast" object passed by the moon again two weeks later. It flew just under the moon to the right, then disappeared behind a mountain.

第一次，「屯門」田景邨
冬天晚上，12am-1am，當時我讀中二，14 歲，約 25 年前
因為好清楚見到個「控」喺月亮前面飛過，佢用⌀呢個軌跡同月亮曾經重疊過，重疊嗰一刻係清楚見到月亮。喺飛過月亮之後消失咗。

第二次，同一個地方屯門田景邨。
冬天晚上，時間真係唔記得，日子大約係第一次見到之後大概 2 星期。

同第一次同一個角度，因為我攤喺張床上睇書，望右邊就係好大隻窗，有咩發光飛過我都好清楚見到。「綠色，類似卡通片入面嘅『激光』」喺天空左邊出現，直線喺月亮底飛過，直到右邊山後就俾座山擋咗，再見唔到。

自從第二次見到之後，我一直都冇打開過窗簾，一年 365 日夜都冇打開過，而家大咗，知道自己見到嘅係咩，當然就咁驚啦，反而望天多咗，好鍾意望天，租樓買樓嘅條件都係要高層，要望到半邊天嘅先會揀，同細個正好相反......最後希望要匿名，唔想畀人當我發神經，感謝。

Case 026

Date: Unknown
Time: Unknown
Location: Unknown
Description (Above): A witness saw a group of formation objects with blinking lights that were not so bright but seemed covered with mist.
Description (Below): The same witness also saw an oval-shaped object with silver/white light. It is surrounded by very bright white light and travels very fast.

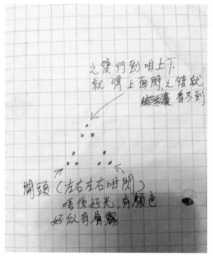

（下）開頭（左右左右咁閃），唔係好
　　　光，有顏色，好似有層霧
（上）之後行到咁上下，就得上面閃，
　　　之後就看不到

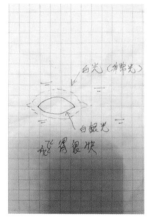

（上）白光（非常光）
（中）白銀光
（下）飛得很快

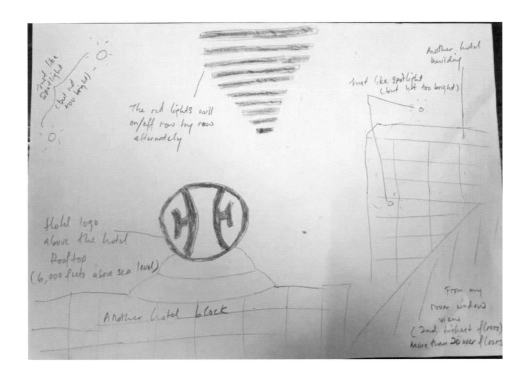

Case 027

(from left to right)
Just like Spotlight (but not too bright) / The red lights will on/off row by row alternately / Just like Spotlight (but not too bright) /Another hotel building

(bottom left)
Hotel logo above the hotel / Pooftop (6,000 feets above sea level)

(bottom right)
From my room window view (2nd highend floors). More than 20 over floors

Structure Crafts

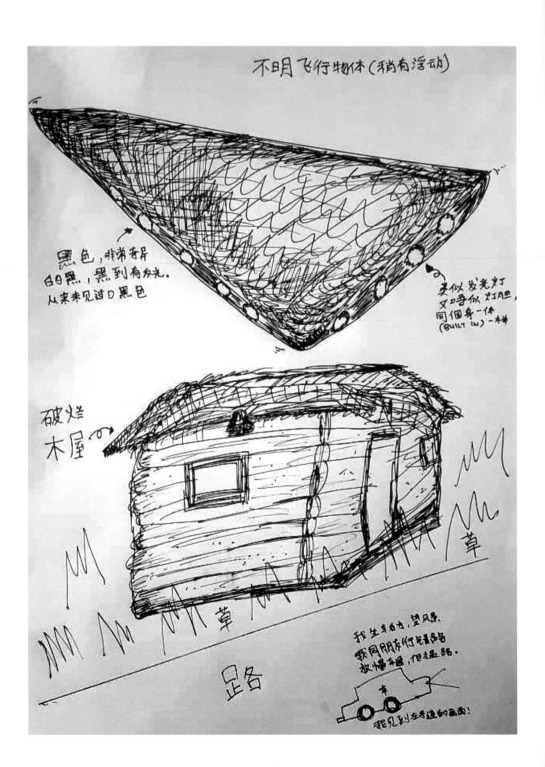

不明飞行物体 (稍有浮动)

黑色,非常奇异
白日黑,黑到有幻光。
从来未见过口黑色

类似发光灯
又唔似灯胆,
同个身一体
(BUILT IN)一样

破烂
木屋

草

草

路

我生本能力,望风景,
我同阴友行见得多,
放慢车速,但无法。

我见到左手边的画面!

64

Case 028

Date: Around 2006
Time: 18:00
Location: Kuala Lumpur, Malaysia
Description:

Three young women were driving to dinner in Kuala Lumpur. Then they get lost in a remote village. The woman at the back saw a big black triangle hovering above one of the wooden houses. The detailed description of the object is as follows: Black but shiny (witness mention she never saw this kind of black color before), some circle-shaped light on the sides but for different with regular lightbulbs. Because they stop the car looking for the right way to get out of the village, the sighting duration lasts a minute. They realize they are experiencing "Missing Time." They believe they spent twenty minutes in the village, but they arrived three hours later.

Ok，本來都唔想提起，因為沒太多朋友相信，朋友都話我亂講話我痴線。但係埋藏心裡多年仍然歷歷在目！

呢件事發生大約 15 年前，當時還是大學生，依家都 3 張幾了。

當時約咗一班同學去某間飯店聚會，大家唔同地點出發，大約傍晚 6pm 一名女同學有車，我就跟咗佢車，仲有一名女同學坐副司機位，我坐後方，聽聞飯店地點在吉隆坡當時舊蘇邦機場附近，但係去到附近找唔到，當時學生時代用啲手機都係幾好嘅，GPS 亦未好完整，後來 call 朋友先知唔係嗰個地點，完全行錯咗路。

結果我地兜兜下行入咗一個偏僻的村子裡。我地完全冇方向感，非常驚慌，於是放慢車速慢慢搵出路。

當時我無車都係路癡亦幫不上忙，就一直望住窗外風景。當去到一段路我見到路旁有間破爛木屋，我地距離係好近，正正木屋就在路邊。女司機朋友停咗一下再打電話問朋友方向，而我再望屋頂上覺得怪怪地，突然嚇到！我見到有一架比木屋仲大 2-3 倍的黑色不明飛行物體，懸浮在木屋屋頂上方！形狀係長方三角形，佢嘅黑色係黑到反光，我從來未見過地球上有呢種顏色。而佢嘅底部邊邊好似有 built in 嘅發光圓圈。當時我係嚇到傻晒，完全想喊都卡著叫不出。因為有停低，我前後都望咗有大約成分鐘，但係真係反應不過來，也沒想到要拍照，畢竟當年我手機差亦有拍照功能，只係 O 住個嘴睇到傻晒。因為朋友 call 緊問路，我都唔好意思打擾叫她睇，腦海有很多想法，又害怕會被攻擊等等畫面，畢竟當時我對神秘學 UFO 呢啲嘢毫無研究。

後尾朋友就開車離開，之後我馬上問佢哋。司機朋友問路冇留意，而最氣死我係坐嘅朋友，佢話有睇到，但係佢係一個無鬼神論者，亦完全唔信有 UFO！佢話係機場嘅展示品或者是氣球，真係氣死我，明明係目擊證人但係竟然否定我。我想馬來西亞當時應該冇可能有咁先進嘅科技有呢種設計，就算依家都唔太可能，唔似係人類的飛行物體。何況係浮動，又點會在木屋上面盤旋。

事後亦都很奇怪，後尾確認地點我哋感覺用咗 20 分鐘就去到正確目的地，結果遲晒大到，大約 9pm 先到達。當時亦不以為意點解個天色黑得咁快，後尾想想覺得奇怪，仲俾班一早就到嘅同學責怪話我哋遲到太多。好奇怪好似你哋所謂嘅 missing time，但係完全冇發覺任何事情，就係突然天色黑晒，遲大到。當時亦冇想太多，而事後亦冇其他變化，一切如常生活。近呢幾年才開始聽神秘學才想起呢件事。

畫畫不是很好，記憶中大概係咁樣嘅畫面。

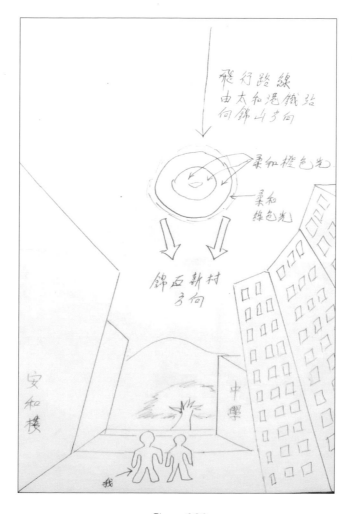

Case 029

Date: 2015
Time: 00:30
Location: Tai Po (Tai Wo Estate), Hong Kong.
Description: Witness saw a big circular object fly in the clear night sky. The bottom of the object grew soft amber light and soft green light on its rim and made green light trails. It was silent when it flew up 40 stories high, and it was as big as the flat surface of the building roof.

（上）飛行路線由太和港鐵站向錦山方向
（中）柔和橙色光／柔和綠色光
（下）錦石新村方向

我名叫 Gary！於 2015 年某月某日夜晚零晨 12 時 30 分，我與朋友由大埔太和邨安和樓地下向錦石新村方向步行，我們左手面是安和樓右手面是寶雅苑（大約兩面樓都有 35 層左右或更高），前方有一所中學！不知為何當晚只是半夜時間不是太夜但條街不多人經過，突然我們頭頂上後方有一隻大型 UFO 向錦石新村及錦山方向高速飛過，只用 3 秒鐘已飛入錦山消失了！我們二人都見到飛行物消失後互望對方，朋友問我是什麼東西？我立刻回答他「UFO」！

因為 UFO 都頗大，當時飛行高度我估有 40 層樓以上，但有一棟大廈天台大小，如降落地面肯定很大隻，它為圓形只見底部，底部全部是發出柔和橙色光但圓邊位全部圍著柔和綠色光，有些拖尾綠光噴出來！它飛行時沒有聲音很靜！不是航拍機！圖畫由我親手畫的但比例大小畫不好！當晚天色很好沒有雲

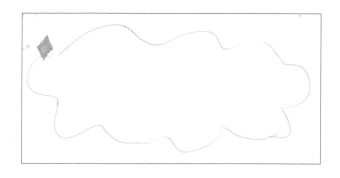

Case 030

Date: 2014 - 2016
Time: 08:30
Location: Exit of Western Harbour Tunnel, Hong Kong.
Description: Witness on a bus and saw a gyro-shaped object station in the sky.

年份：5、6 年前夏天
地點：西隧出入口 (本人正乘搭 948 號巴士)
時間：上午 8:30
當巴士駛近西隧口時，我好奇望住個天，見到有一粒好似「陀螺」物體動也不動，好似「痴」住嚿雲咁停咗响度，到我由西隧出番嚟時佢都冇移動過。

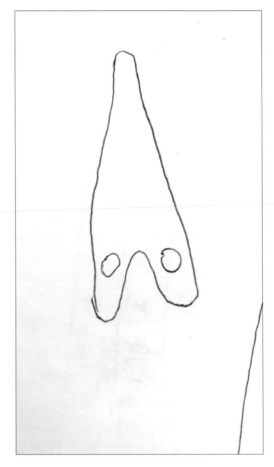

Case 031

Date: Unknown

Time: Unknown

Description: No written description from the witness.

日期：不明

地點：不明

描述：目擊者並未提供文字描述

Case 032

Date: 25 August 2003
Time: Around sunset (Not yet dark)
Location: Hong Kong Island East
Description: A diamond-shaped object with blue light and trails flew across the harbor. It passed to West from East.

2003 月 8 月 25 號，響日落但係天未黑齊嘅時候，見到呢一隻 UFO，向天上面掠過。

佢飛行嘅時候冇聲，發出藍色嘅光，甚至有慧星嘅尾巴，我當時在香港島東區，背向九龍面向南方，呢隻物體有我嘅視線入面，由東向西飛過。

點解咁肯定係嗰一日，因為目擊呢隻 UFO 之後過咗兩日，大嶼山有直升機墜毀嘅意外，所以特別記得個日子。

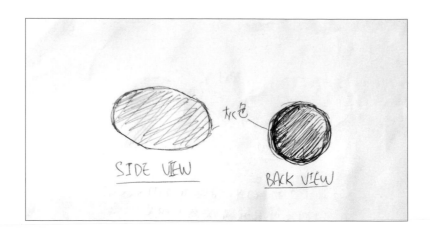

灰色

SIDE VIEW

BACK VIEW

Case 033

Date: Early October 2012
Time: Around 08:00
Location: Tuen Mun (King Fung Playground), Hong Kong.
Description: An oval-shaped object slowly flying in the sky, the sighting lasts two to three minutes.

2012 年 10 月初，早上約 8 時。當時帶緊小朋友返學，行到屯門景峰籃球場，突然抬高頭見到遠處有嚿圓形拉長咗嘅灰色嘢喺天度慢飛，我小朋友話見唔到。我哋一路行，一路睇住佢慢飛，小朋友仍然話見唔到。街上有好多人，唔知啲人係睇唔到，定冇留意，當時係得我望天。行咗 2 至 3 分鐘，送咗小朋友返學，我走番行人路，仍然見到佢喺遠處飛緊，當時係見到佢後面，係一個圓球形。佢一路飛，我一路跟佢尾行，直到飛過咗山頭，睇佢唔到。

Case 034

Date: Unknown
Time: Unknown
Description: No written description from the witness.

日期：不明
地點：不明
描述：目擊者並未提供文字描述

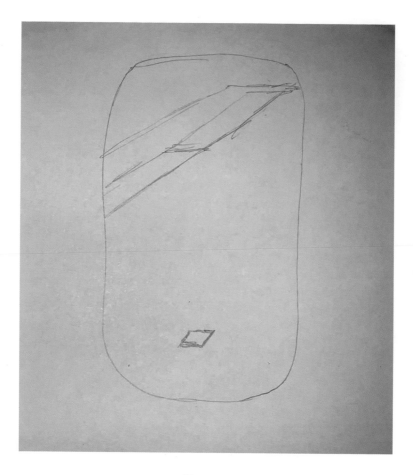

Case 035

Date: December 2012
Time: 10:30 - 11:30
Location: Airplane to Thailand
Description: Four passengers saw a regular-shaped object flying in the air next to the airplane. It looks like a piece of A4 paper. It keeps the speed with the plane, disappearing five minutes later.

2012 年 12 月某日上午 1030-1130 之間
在飛機往泰國途中在窗邊向下望,見有一張金屬銀色物體停在空中。四方形感覺好像 A4 紙飄在空中,一路同步飛行,大約 5 分鐘後消失。同場有 4 個伙伴睇到。

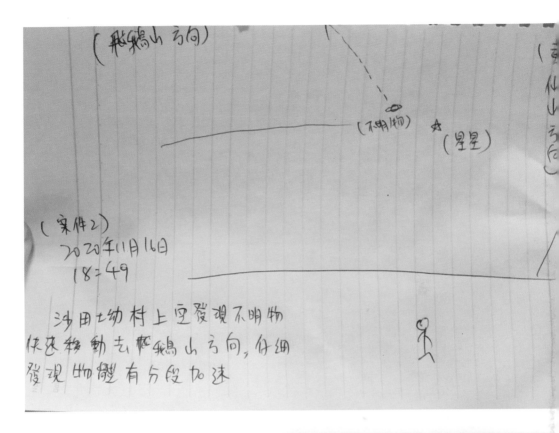

（飛鵝山方向）

（不明物）

☆（星星）

（黃仙山方向）

（案件2）
2020年11月16日
18:49

沙田坳村上空發現不明物
快速移動去飛鵝山方向，仔細
發現物體有分段加速

Case 036

Date: 16 November 2020
Time: 18:49
Location: Tsz Wan Shan (Shatin Pass Estate), Hong Kong.
Description: Witness saw a Saturn-shaped object flying in the sky at high speed. Witness describes it flew with accelerated speed.

2020 年 11 月 16 日 18:49
沙田坳村上空發現不明物快速移動去飛鵝山方向，仔細發現物體有分段加速

蓄規泡外圍的線像火光?↓

UFO 雪茄型:

表面光滑. 無窗. 無噴射器
等等. 看似石膏表面!
平滑. 不似 Youtube.

蓄規泡表面七彩色

外圍有紅色邊

Case 037

Date: Around 1998
Time: After 17:00 (Sunset time)
Location: Exit of Shing Mun Tunnel, Tsuen Wan, Hong Kong.
Description: A taxi driver and passenger saw a black cigar-shaped object hovering in the sky. A circular colorful "Bubble" surrounded it with a red edge outside. Amber or colorful grows outside the circular shape. "Very beautiful! I can never forget that day, and so relieved now after I said it publicly." the taxi driver said.

MJ13 你好,我係亞倫,詳細講一講。我二十二年前做的士司機,當時我 22 歲至 23 歲,剛考到的士牌,我的身體和雙眼正常的。當日大約下午五時後,我在沙田新城市廣場接了一位男士上車,他要出荃灣。我由城門隧道過了收費亭,大約左邊是巴士轉車站尾,這位乘客突然大叫「司機,司機見唔見到呀!」我被佢嚇一嚇,我自然反應望向左邊,只見到有很多人等巴士,這位男乘客突然又用手指指向我前面右邊天上,佢好大聲問我「咩嚟㗎?」,我當時望到,心裡嘩了一聲又再嚇一嚇!我將部的士煞停,係急煞停車,我見到天上有條炭灰黑色似粉筆的東西有番梘泡包住停留在天上,我周圍望有沒有吊車吊起它?它怎會停留在天上!當時我有想過是人造衛星嗎?但這東西左右兩邊沒太陽能板,我再望支粉筆下面沒有噴射氣,粉筆頭我睇唔清楚因為它向後側,另外令我非常深刻的是包住它的番梘泡,好靚,因為在黃昏,所以番梘泡發出光線和彩色光,在外面有橙黃色像極光一樣包住番梘泡慢慢搖動。後面位乘客不停拍櫈,我好像醒一醒,他話趕時間不要再睇,所以我開車走咗。本想趕他落車的,我做唔出,佢大隻過我,佢年紀應該當時大約三十幾四十歲到,希望佢現在聽到都回應吓。多謝 MJ13 可以給我用文字記錄下來!多謝,我心裡舒服多了,二十二年了。

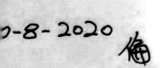

2018年某日黃昏時段 屯門
屯興路望向西南方
當時有一隻小的橙色長型
不明物體 懸浮於空中
當時天色昏暗,未能看
清不明物體.
Claudia Chan

Case 038

Date: 2018
Time: Around sunset time
Location: Tuen Mun (Tuen Hing Road), Hong Kong.
Description: A tiny orange cigar-shaped object hovering in the sky. "It was dark, so I can't see it clearly."

2018 年某日黃昏時段屯門屯興路望向西南方。

當時有一很小的橙色長型不明物體懸浮於天空中，當時天色昏暗，未能看清不明物體。

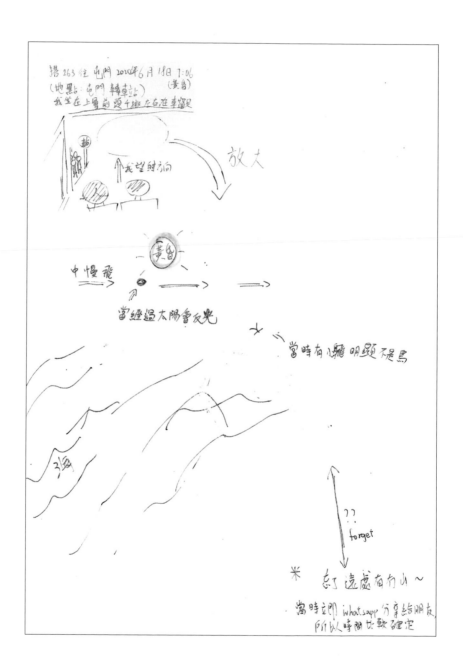

Case 039

Date: 18 June 2020
Time: 19:06
Location: Tuen Mun traffic interchange station, Hong Kong.
Description: An oval-shaped object flew slowly in the sky. Witness ensure it was not a bird.

（上）搭 263 往屯門 2020 年 6 月 18 日 黃昏 7:06（地點：屯門轉車站）我坐在上層前頭 4
　　　排左右，在車窗見
（中）中慢飛 / 當經過太陽會反光 / 當時有小鳥飛　明顯不是鳥
（下）忘了遠處有冇山 / 當時立即 Whatsapp 分享給朋友，所以時間比較確定

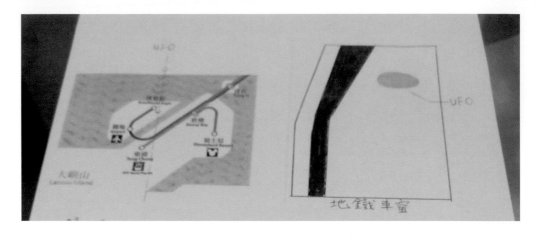

Case 040

Date: 15 July 2017
Time: Before 13:00
Location: Train to Internation Airport from Tsing Yi, Hong Kong.
Description: A family of four witnessed an egged-shaped object with a smooth metallic surface. They found no windows or wings, and it had no lights, just like a big "Silver egg" hovering in the sky. The sighting lasts for about 30 seconds until their train turns in a different direction. They have cell phones, but the sighting makes them forget about filming it.

目擊日期：15.7.2017
時間：約中午 1 點前
地點：乘地鐵由青衣往機場博覽館途中
經過：15.7.2017 中午我和家人四人乘地鐵從荔景站轉車往博覽站參觀亞洲國際博覽館所舉辦的爬蟲展。

　　當列車駛經青衣至東涌其間，我發現有飛行物從橫向飛過列車，起初從遠處飛來時以為是看到爬升中的飛機，但越睇越懷疑，因為根本見唔到機翼！

　　當日是中午，天氣晴朗，萬里無雲，因此視野非常清晰，飛行物是以慢速直線橫飛過列車，因此從我們角度就似看到飛機爬升中，為了確認我所看到的東西，還叫了兒子及女兒幫忙觀察，我們三人貼在列車車窗觀看著，而太太因位置問題並未在意我們。

　　經過差不多半分鐘觀察，我們都意識到那個不是飛機，近看比飛機細及短，那是一個全銀色蛋型 UFO，表面光滑，沒有窗，沒有翼，沒有信號燈或發出光，純焠是一個金屬銀色的大雞蛋在天空中飛！

　　最後因列車轉彎而無法睇到。當日我們是有帶手機的，但因為太專注觀看，根本冇意識到去影相直至飛過咗先記起，可惜！

　　太興奮了竟然看到了 UFO，哈哈！29.10.2020

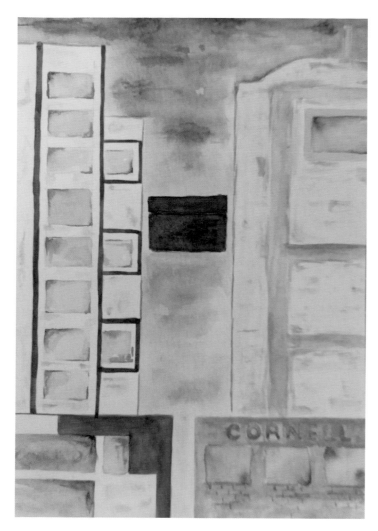

Case 041

Date: November 2016
Time: Around 23:30
Location: Unknown
Description: A cargo-shaped object flew between two industrial buildings, and it passed smoothly at around thirteen to fifteen stories high. The top part of the object is red, and the bottom part is blue.

2016 年 11 月晚上大概 11:30
望向柴灣避風塘位置，喺 2 幢工業大廈之間，見到一隻貨櫃般大嘅 UFO 喺 2 幢大廈之間飛過，UFO 上面部份紅色下部份藍色，高度大概半幢樓高（13-15 層）好平穩飛過。

Case 042

Date: 24 September 2019
Time: 18:17
Location: Ferry from Kwun Tong to Sai Wan Ho, Hong
Kong harbor.
Description: A shooting-star-shaped (Teardrop-shaped)
object flew slowly to the Peak.

2019 年 9 月 24 日黃昏 6:17
由觀塘碼頭搭船到西灣河，經過郵輪碼頭時，望
見郵輪碼頭上空，有 1 粒好似流星型態嘅 UFO，
由郵輪碼頭位置，一直慢飛到中環太平山方向，
直到離開我視線範圍。

Case 043

Date: End of September 2020
Time: 09:30
Location: Unknown, Hong Kong.
Description: A white circle-shaped object flew into the cloud. The witness saw it in her bedroom.

2020 年 9 月尾早上 9:30

在睡房開窗，抬頭望，便見到一粒白色圓形 UFO 飛入雲中。

Grace Fok

NOV.2020

Case 044

Date: 25 September 2017 (Monday)
Time: 15:30
Location: Kwun Tong (Industrial area), Hong Kong.
Description: A boy witnessed a Saturn-shaped object hovering in the sky, wafting high up. The top and bottom parts are translucent domes, and the rim in the middle grows white light. Flowing sometime in the sky and suddenly flying away with high speed.

給你一些資料作參考。雖然小孩的話未能盡信,怕他表達有誤差,但因他從自細到大也聽我說很多也看過一些網上的片段,所以略知一二。應分得出咩唔係飛機 😜 那天小兒在校巴內望天時見到疑似 UFO,外形如下圖(他畫的)

時間:25 sept 2017(星期一),3:30 左右
地點:觀塘,工廠區,美亞廚具那組大廈的上空
形狀:如下圖,中間是發光(白光)的長條形。上下有半透明的半圓形。
尺寸:由小兒說,它的位置比一般飛機高。中間一條的闊度約兩隻手指粗(估計 2cm 左右),長度約為一塊大薯片(估計約 7cm 左右)。而上下半透明部分的闊度和中間部分的差不多(估算全隻闊 6cm)。

當時情況:我仔仔無法說出它在天空停留了多久,只是說開頭慢慢由牛頭角向藍田方向飛,突然加速,便不見了。

Case 045

Date: May 2018
Time: Around sunset
Location: The Peak, Hong Kong.
Description: A sphere object slowly fly by from Mt. David. It was purple with the mirror surface. Sized as big as a van and passed to the west.

2018 年 5 月某一天將近黃昏時段　凌霄閣
一個略帶紫色的鏡面球體，從摩星嶺的左後方緩緩飛出，以忽高忽低但不誇張的幅度，飛過摩星嶺的上空，甚至近距離掠過山上類似電波塔的建築，我就在凌霄閣天台，所以很清楚地判斷該球體大概有一輛 7 人車般大小，掠過摩星嶺後便往西飛走了。

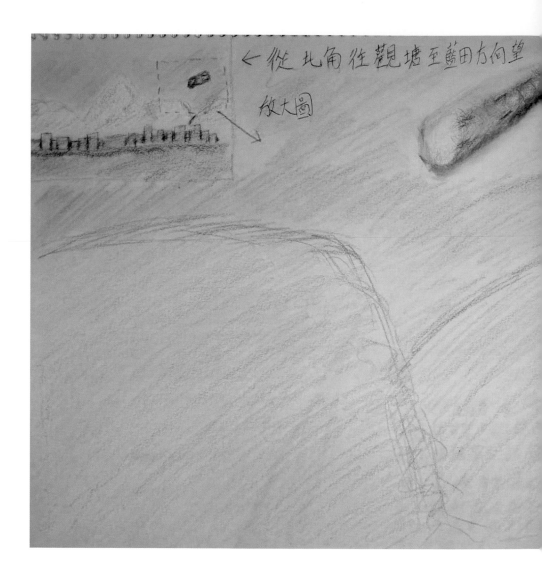

← 從 北角 往 觀塘 至 藍田方向望

放大圖

Case 046

016年1月16日下午
一形的長條狀物體
引在觀塘與藍田
的後方,非常大
幾百米長,可能唔止
正清楚而見那東西的
在,但眼睛就是
法聚焦到它上

時我雙眼看其他
物都可正常聚焦
獨該物體怎樣
它,它都像對焦
敗的攝影作品
樣,是 out fo 的

Date: 16 January 2016
Time: Afternoon
Location: Kwun Tong / Nam Tin, Hong Kong.
Description: A Cigar-shaped object hovering in the sky; it was massive, at least a few hundred meters long, and hovering above Kwun Tong or Nam Tin. Witness mentioned it looked "Out of focus," it was blurry but solid.

2016 年 1 月 16 日下午

雪卡形的長條狀物體,目測在觀塘與藍田山上的後方,非常大,至少幾百米長,可能唔止。

＊很清楚看見那東西的存在,但眼睛就是無法聚焦到它上。

當時我雙眼看其他景物都可正當聚焦,唯獨該物體怎樣看它,它都像對焦失敗的攝影作品一樣,是 out fo 的。

Case 047

Date: 14 August 2004
Time: 22:10
Location: Lei Yu Mun, Hong Kong.
Description: A sport car-shaped object surrounded by
green grows appeared shortly before the fishing witness.
The object appeared one or two seconds before vanishing
in the empty air. It is a single witness case.

事發：喺雲層間出現一架被一團綠光包圍嘅跑車
　　　形 UFO，團綠光 UFO 只維持一、兩秒就
　　　消失咗。因為當時釣緊魚，正岳高頭睩實
　　　枝魚竿，所以雖然只出現一兩秒，但都睇
　　　得非常清楚。(可惜當時只得我一個在現
　　　場)
時間：2004 年 8 月 14 日約晚上 10 時 10 分
地點：鯉魚門「大佛口食坑」旁邊嘅大平台 (現為
　　　「南大門」酒家)

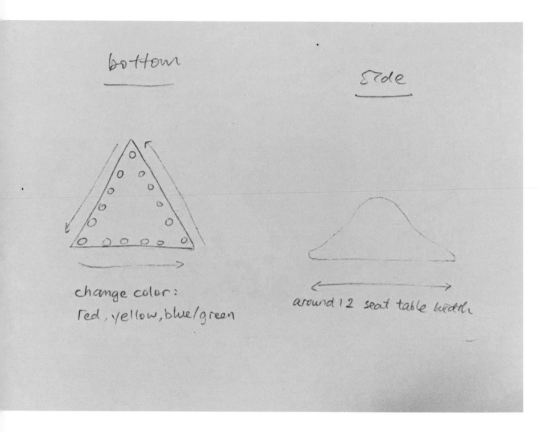

bottom

side

change color:
red, yellow, blue/green

around 12 seat table width

Case 048

Date: 23 December 2012
Time: 22:00 or after
Location: Sai Kung, Hong Kong.
Description: A triangular object appeared above the roof of the witness's home, and it was flashing colorful lights on the bottom, moving around up the top of the witness' house. It was as wide as a twelve-seat table. The lights on the bottom keep changing from red to yellow, then blue to green.

日期：2012 年 12 月 23 日

時間：10:00pm 或之後

地點：香港西貢

描述：一個三角形的物體出現目擊者家的屋頂上方。它底部閃爍著五顏六色的燈光，不斷
地從紅色變為黃色，然後從藍色變為綠色。它有十二個座位的桌子那麼寬，在屋頂
周圍移動。

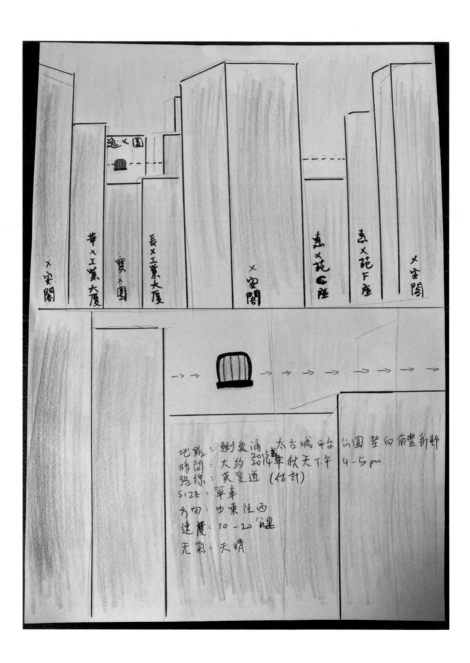

地點：鰂魚涌 太古城平台 公園望向南豐新邨
時間：大約 2014年秋天下午 4-5 pm
路線：英皇道 (估計)
SIZE：單車
方向：由東往西
速度：10-20 哩
天氣：天晴

Case 049

Date: Autumn of 2013 or 2014
Time: 16:00 - 17:00
Location: Tai Koo Shing, Quarry Bay, Hong Kong.
Description: An odd object flew from East to West about 10-20km/h. It was as big as a bicycle. It was a clear day.

地點：鰂魚涌太古城平台公園望向南豐新邨
時間：大約 2013 年或 2014 年秋天下午 4-5pm
路線：英皇道（估計）
SIZE：單車
方向：由東往西
速度：10-20 公里
天氣：天晴

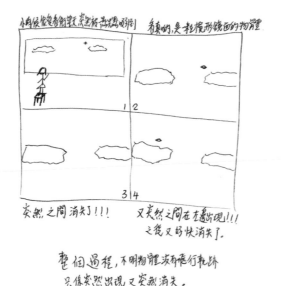

Case 050

Date: Around the summer of 1995
Time: 17:00 - 18:00
Location: Choi Wan Estate, Wong Tai Sin, Hong Kong.
Description: A rugby-shaped object with a mirror surface appeared in the sky. Then it disappears suddenly and reappears on the left side from the original position. Then it disappears again. "The object didn't fly; it appeared and disappeared around the sky!" the witness explained.

時間：大概 1995 年 (夏天) 下午五至六時左右
地點：彩雲邨，甘霖樓 (高層) 望向坪石遊樂場的上空位置

1 小時候很愛看天，突然被一點光點吸引到
2 看真啲，是一粒欖形鏡面的物體
3 突然之間消失了！！！
4 又突然之間在左邊出現！！！之後又好快消失了。
整個過程，不明物體沒有飛行軌跡。只係突然出現又突然消失。

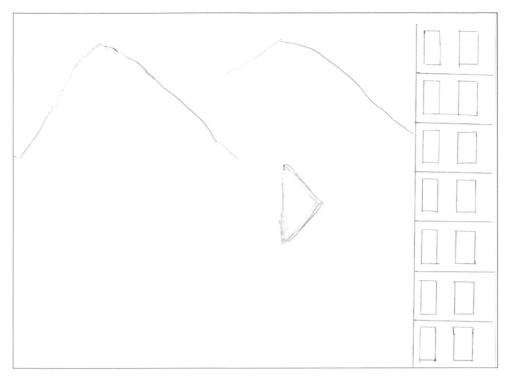

Case 051

Date: Around 1993 - 1995 (Not winter for sure)
Time: Afternoon
Location: Tuen Mun, Hong Kong.
Description: A white translucent object flew from left to right and disappeared behind the building. The translucent surface is not even, and some areas are more white than the others. (Witness mentioned that the British army base on top of the hill should still be operating back then.)

位置：屯門良景
時間：約 1993 - 1995 年，非冬季，下午
事發經過：當時讀小學，下午在房間望出窗外，見到一個白色半透明三角形，不規則地有些位置比較透明，有些位置比較白色，由左至右，飛到隔籬大樓後面，被大樓遮擋再無法看見，山上有英軍基地，當時基地應該有在使用。

99

Case 052

Date: 26 July 2012
Time: 17:00 - 02:00 next day
Location: Shanghai, China.
Description: A trapezoid blue light appeared in the sky, then three growing objects appeared and spun around the light. They occasionally make zip-zap turns or station in the sky.
Witness observes the sighting from 30/Floors building; they flew closer at later hours of the night. At the end of the sighting, a piece of fire arrived from the sky above and stopped on top of the opposite building. It suddenly makes a 90-degree turn and is surrounded by white smoke.

Shanghai 26 Jul 2012

從下午 5 點至凌晨 2 點

突然出現藍梯型光在天空，於是用 ipad 拍片，慢慢發現有三個發光物。

邊發白光，邊打訊號，仲會變紅色，物件放大看會上下左右來回轉，還會左右兩邊 Z 型的軌跡移動及停頓。我從 30 樓酒店望出去，到夜晚越飛越近。見下一張。

（在同一房間看出去）

到第二天早上，看見一團煙火加大火光從天空向下衝，衝到大廈頂，突然煞停，然後 90 度向前飛，完全被團白煙包住。

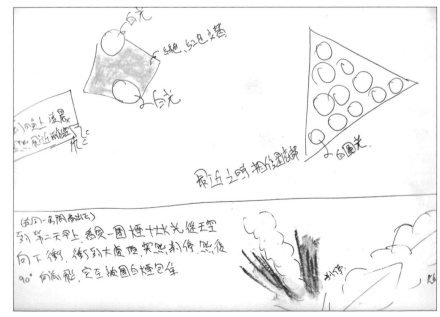

101

Case 053

Date: Around 1971
Time: Evening
Location: Kwun Tong, Hong Kong.
Description: A group of children witnessed a disc-shaped object while playing on the public building roof. The machine's noise caught their attention at first, and then they saw the spinning bottom of the disc grow red light. Witness ensure it was neither helicopter nor airplane. He never mentioned it to anyone but talked to two brothers to confirm the sighting.

應該是 50 年前的事件，當時不到 10 歲，我們住在現今玉蓮台所在的廉租屋。當晚我如常與一班鄰居小朋友在天檯玩耍，突然從後面天空傳來連續不斷的機器嘈雜聲。我立即從左邊向後望，見到一隻圓碟形的東西從身後的天空（那邊是觀塘牛頭角對開的海面，當年工廠大廈也不多）一直從左邊天空經過頭頂再飛向右上方（應該是鱷魚山的右邊方向）。那隻圓碟形東西底部是會旋轉的並發出紅光，而光線像是從旋轉的窗門射出。我肯定這不是直昇機，更不是飛機，它的體積比直昇機大，速度較直昇機快，聲響與直昇機大不相同，飛行高度比直昇機高。就這次而已，亦沒有再跟其他人提及；數十年過後，偶然跟我的兩個弟弟訴說當晚情況，豈料他們當時也看到那隻飛碟，令我更確定自己所見到的是真實的事情。

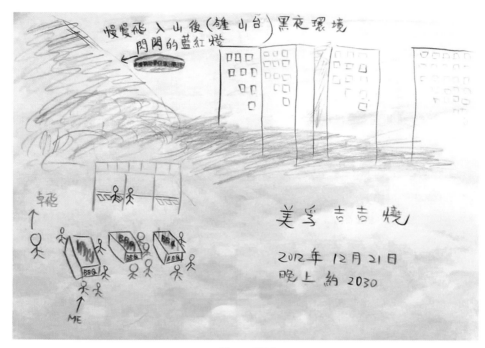

慢慢飛入山後(鐘山台)黑夜環境
閃閃的藍紅燈

美孚 吉吉燒

2012年 12月21日
晚上 約 2030

卓飛
↑
人

ME

Case 054

Date: 21 December 2012
Time: Around 20:30
Location: Mei Fu, Hong Kong.
Description: A large flying object appeared in the sky during a barbeque party. A Bun-shaped flying object appeared with the silverish grey outer shell. It has a red and blue flashing light that reminds a police vehicle. Withness is so exciting, but it seems she is the only one who can witness it. People around her ignored her excitement. An hour later, she decided to snap off a couple of pictures of the sighting area. One of the images shows three small dots on the inverted image.

2012 年 12 月 21 日盛傳這日是末日到來，所以 MJ13 在美孚吉吉燒搞了個「末日BBQ」，希望與 Fans 朋友們一起共渡過末日來臨，大約晚上 8 時半卓飛正在分享對 UFO 事情時，我面向荔欣苑呆呆的望上暗黑天空，突然我見到一架大型飛行物出現，外形好似日本紅豆餅，外殼暗銀灰色，最深刻中間有紅藍色燈閃動，好似警車上警號燈一樣顏色，飛行物幾秒就飛過了，去「鐘山台」後面，根本趕不及取相機或者手機影相，飛行物出現的一刻，我有即時彈起身，但好似沒有人有反應！？好似我與其他人不同，莫非只有我看到！？

我繼續坐下，卓飛繼續分享，其他人繼續燒烤，但這一刻我好興奮，第一次見到就是大型UFO，早在卓飛宣佈搞「末日BBQ」時，我已經每日都講出聲：「我要睇到 UFO！我要睇到 UFO！我要睇到 UFO！」直至 BBQ Party 當日！其實我好想同他說出剛才我見到什麼，由於當時未熟識他，怕他不相信，怕其他人不相信，所以就把事情收起。

大約一小時後，已經有 Fans 排隊向卓飛索取簽名，我無聊時向樂園 A 座對出天空影了張相，心理上覺得剛才沒有影到 UFO，現在補影回末日晚的天空作個紀念，第一張相是有排燈光，後來才知道，這是取食物位置的一個簷篷下的一排燈泡燈光反影，我覺得好奇怪，再影第二次，是大廈上空，沒有比較。聽多了 MJ13 節目，得知有時同場，未必個個都能見到 UFO，可能有些人見到，有些見不到，及後傳送兩張相給 Stan Ho 看過，他認為第一張相是排燈泡燈光反影，Stan 把第二張相反白了，竟然見到有奇怪 3 點出現，普通相是見不到，只反白才見到，同一晚發生兩次奇遇？！

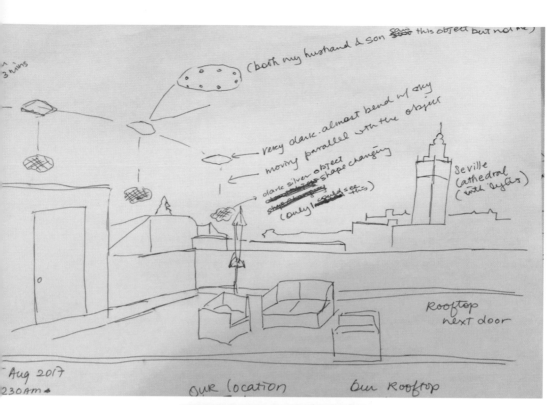

Case 055

Date: 5 August 2017

Time: 00:00 - 00:30

Location: Unknown

Description:

-(both my husband & son saw this object but not me)

-very classic almost bend of sky

-moving parallel with the object

-dark silver object

-shape changing (only 1 could see this)

日期：2017 年 8 月 5 日

時間：00:00 - 00:30

地點：不明

描述：

-（我的丈夫和兒子都看到了這物體，但我沒有）

- 非常經典天空幾乎扭曲

- 與物體平行移動

- 暗銀色物體

- 形狀變化（只得 1 人見到）

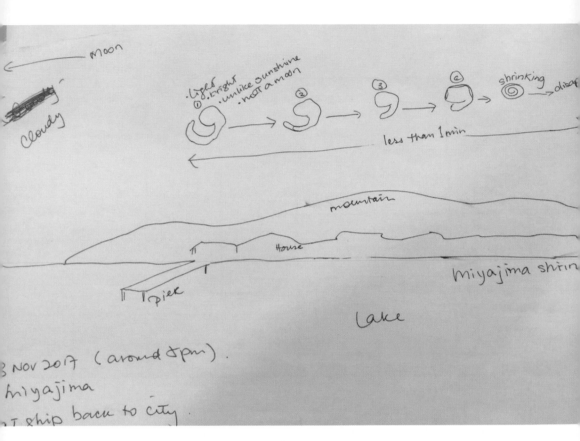

Case 056

Date: 3 November 2017
Time: Around 05:00
Location: Miyajima (at ship bacu to city)
Description:
-light
-bright
-unlike sunshine
-not a moon
-shrinking->disappear
-less than 1 min

日期：2017 年 11 月 3 日
時間：05:00 左右
地點：宮島
描述：
- 光
- 明亮的
- 不像陽光
- 不是月亮
- 收縮→消失
- 少於 1 分鐘

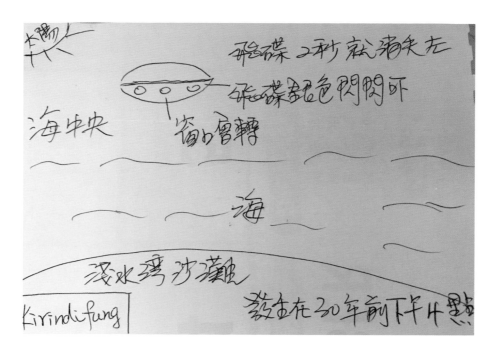

Case 057

Date: Around 1992
Time: 16:00
Location: Repulse Bay Beach, Hong Kong.
Description: A flying saucer with a silvery flash appeared in the sky, then vanished two seconds later. Withness claimed three "Windows" are"rotating."

淺水灣沙灘。發生在 30 年前下午 4 點
-海中央
-飛碟 2 秒就消失咗
-飛碟銀色閃閃吓
-窗口會轉

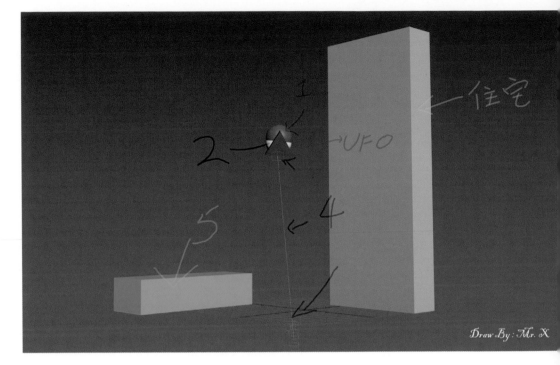

Case 058

Date: Unknown
Time: Unknown
Description: No written description from the witness.

日期：不明
地點：不明
描述：目擊者並未提供文字描述

Case 059

Date: Unknown
Location: Kwai Chung Container Terminal, Hong Kong.
Description: A strange object that floated in the sky.

日期：不明
地點：香港葵涌貨櫃碼頭。
描述：一個漂浮在天空中的奇怪物體。

Encounter

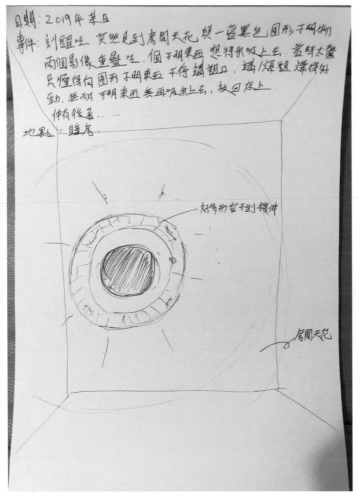

日期：2019 年某日
事件：瞓醒咗，突然見到房間天花與一嚿黑色圓形不明物
兩個影像重疊咗，個不明東西想特我吸上去，當時太驚
只懂得向圓形不明東西干停講粗口。講／爆粗爆得好
勁，無耐不明東西無再吸我上去，放回床上。
仲有後著……
地點：睡房

好像形容干到撐件
房間天花

Case 060a

Date: One day in 2019
Time: Unknown
Location: Witness' bedroom, somewhere in Hong Kong.
Description: Wake up at an unknown time, and the witness claimed an image of a circular-shaped object appeared on the ceiling. And then have a feeling of being pulled up to it. The witness feels very frightened and starts to curse, like REAL curse :-D (People in Hong Kong intended to curse or scold to get rid of paranormal activity when they are terrified!)

The curse seems to be working well, then the pulling stop...

日期：2019 年某日
事件：瞓醒咗，突然見到房間天花與一嚿黑色圓形不明物兩個影像重疊咗，個不明東西想將我吸上去，當時太驚只懂得向圓形不明東西講粗口，講／爆粗爆得好勁，無耐不明東西無再吸我上去，放回床上。仲有後著……
地點：睡房

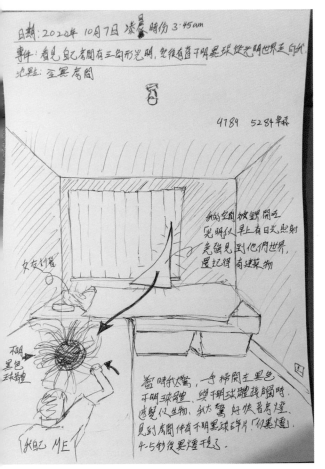

Case 060b

Date: 7 October 2020
Time: 03:45
Location: Witness' bedroom, somewhere in Hong Kong.
Description: The Witness claimed something "Slid open" his "space." And intense light shined in through the "slid." He remembers he saw "buildings" through the "slid."
He then saw a "black sphere" run towards him, and he pushed it to a side as he was so terrified. The "black sphere" feels like a living entity when touched.

Witness turned on the light in the room immediately. A black circular-shaped of fragments still can be seen. At the same time, they were fading out like smoke in 4 to 5 seconds when the light was on.

日期：2020 年 10 月 7 日 凌晨時份 3:45am
事件：看見自己房間有三角形光明，然後有嚐不明黑球從光明世界走向我。
地點：全黑房間
（中）我的空間被鉾開咗。「光明」似早上有日光照射，勉強見到他們世界，還記得有建築物
（下）當時我太驚，一手掃開咗黑色不明球體，與不明球體接觸時，感覺似生物，我太驚好快著房燈，見到房間仲有不明黑球碎片，「似黑煙」，4-5秒後黑煙不見了。

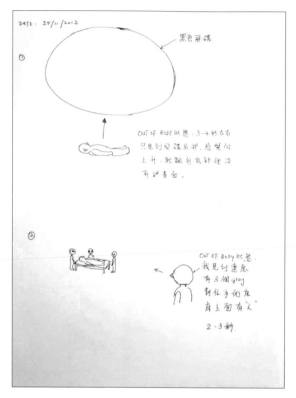

Case 061

Date: 25 November 2012
Time: Unknown
Location: Unknown
Description:

1. Black UFO / Out of body experience state, 3 to 4 seconds, only saw the bottom part of the flying saucer, have a feeling of pulling up, blackout when close to the bottom.
2. Out of body experience, the witness saw three grey aliens surrounding the bed; a "person" was on the bed for 2 to 3 seconds.

   The witness went to hypnotize regression in December of 2018. Saw three grey aliens and realized he was lying on a bed with lights on top of his head. Three grey aliens were wearing a turtle neck tide fitted black suit standing on his left side. Then one of them walked to his right side. That's all the memories. The whole memory fragment lasts one to two minutes long.

   The witness went to another regression on 7 January 2020. He recalled himself on another planet. The sky and land are in orange color. A white flying saucer is in front of him, and a grey alien is standing next. He remembers walking up the stairs that connect to the bottom of the white flying saucer but cannot get in.

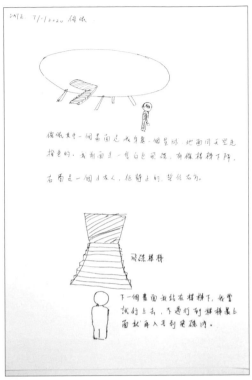

日期：25/11/2012

1. 黑色飛碟 / OUT OF BODY 狀態，3-4 秒左右，只見到飛碟底部，感覺向上升，就飄到
   底部便沒有畫面。

2. OUT OF BODY 狀態：我見到遠處有 3 個 grey 對住手術床。床上面有「人」，2-3 秒。

日期：2018 年 12 月做咗一次催眠，催眠下見到 3 個小灰人。

我意識到自己瞓喺一張床上，應該係手術床，頭頂上有燈，左手面原本有 3 個小灰人，穿著
黑色緊身黑色樽領衫，之後有一個行過嚟我右手邊，整個過程大概 1 分鐘至兩分鐘。

日期：7/1/2020催眠

(中)　催眠其中一個畫面是我身處一個星球，地面同天空是橙色的。我前面是一隻白色飛
　　　碟，有條樓梯下降。右面是一個小灰人，佢靜止的，望住右方。

(下)　飛碟樓梯/ 下一個畫面我站在樓梯下，我嘗試行上去，不過行到樓梯最上面就再入唔
　　　到飛碟內。

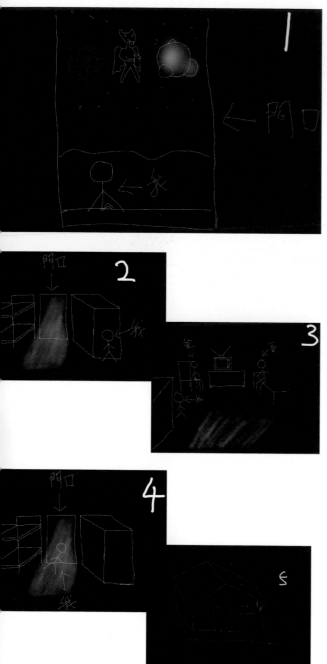

Case 062

Date: Unknown
Time: Unknown
Location: Lo Wu, Sandy Ridge, Hong Kong.
Description:

1. The witness claimed he heard a mechanical sound after dinner; the sound came from outside the house and got louder. He then saw three strange objects floating in the sky. The one on the right will grow light, and it is shaped like the Marshmallow Man in the movie "Ghostbusters." The object in the middle has red eyes, pointed ears, and enormous bat wings on its back. And the object on the left is an unregular shaped object.

2. He was so frightened and ran back into the house, hiding behind a big piece of furniture.

3. He wanted to scream for help but realized he had no sound and people around him were "stop in motion," including the television.

4. He felt being pulled and floating up in the air.

5. He floated up in the air and saw his house in the air.

6. The middle object was catching him.

Witness had no more memory until on bed. Not sure if it was a dream or anything supernatural. He mentioned this to family members, and they believe it was a paranormal experience because they lived close to a cemetery.

1. 我今年 39 歲，我嘅經歷發生喺我三歲。我住羅湖沙嶺近墳場附近。嗰晚我同家人晚飯後一齊睇電視，突然有機器聲音喺屋外邊傳嚟，漸漸變大聲，我行出門口，出現三個物體喺空中。我望出去右邊個物體會發光，以前我覺得係一個白色嘅叮噹，但而家覺得似棉花糖鬼。中間係一個紅色眼、有尖牙、長耳朵。背部有好似翼咁嘅嘢，有啲似一隻大蝙蝠。左邊只係記得暗紅色不規則型態生物。

2. 我好驚，跑返入屋，躲藏喺櫃後面。

3. 想大聲叫家，但係我叫唔出聲，而且電視畫面同屋企人靜止咗。

4. 之後我浮起，我拉住櫃門，但都係拉咗出門口。

5. 升出咗門口，我喺空中睇到間屋頂。

6. 我俾中間人型物體捉住，因為人型物體捉住我係佢咗邊，我記得左邊係白色發光綿花糖鬼光球。右邊物體唔係太記得，只記得係暗紅色。之後就冇記憶，到我有記憶喺床上面。我唔肯定係夢定真係見到怪嘢，如果係夢，到而家都記得其實都覺得有啲奇怪。我同屋企人提過，佢哋覺得我見到鬼，因為有墳場。

UFO Drawings
From the Cantonese world
廣東話世界的不明飛行物繪本

作者　　　：卓飛
出版人　　：Nathan Wong
編輯　　　：尼頓
設計　　　：叉燒飯

出版　　　：筆求人工作室有限公司 Seeker Publication Ltd.
地址　　　：觀塘偉業街189號金寶工業大廈2樓A15室
電郵　　　：penseekerhk@gmail.com
網址　　　：www.seekerpublication.com

發行　　　：泛華發行代理有限公司
地址　　　：香港新界將軍澳工業邨駿昌街七號星島新聞集團大廈
查詢　　　：gccd@singtaonewscorp.com

國際書號　：978-988-75975-3-7
出版日期　：2022年6月
定價　　　：港幣180元(平裝) / 280元(精裝)

Author　　　：Cheuk Fei
Publisher　　：Nathan Wong
Editor　　　：Niton
Designer　　：叉燒飯

Publication　：Seeker Publication Ltd.
Address　　　：Room 15, Unit A, 2/F, Jumbo Industrial Building, 189
　　　　　　　　Wai Yip St., Kwun Tong, Hong Kong
Email　　　　：penseekerhk@gmail.com
URL　　　　　：www.seekerpublication.com

ISBN　　　　　：978-988-75975-3-7
Publcation Date　：June 2022
Price　　　　　：HKD$180 (softcover) / HKD$280 (hardcover)

筆求人
Seeker Publication